SEEDS OF CONCERN

SEEDS OF CONCERN
THE GENETIC MANIPULATION OF PLANTS

David R Murray

A UNSW Press book

Published in Australia, New Zealand,
Papua New Guinea and Oceania by
University of New South Wales Press Ltd
University of New South Wales
UNSW Sydney NSW 2052
AUSTRALIA
www.unswpress.com.au

and in the rest of the world by
CABI Publishing
CAB International
Wallingford, Oxon OX10 8DE, UK
Tel + 44 (0) 1491 832111 Fax + 44 (0) 1491 833508
Email <cabi@cabi.org>

CABI Publishing
10E 40th Street, Suite 3203
New York, NY 10016, USA
Tel + 1 212 481 7018 Fax + 1 212 686 7993
Email<cabi-nao@cabi.org>
Web site: www.cabi-publishing.org

© David R Murray 2003
First published 2003

This book is copyright. Apart from any fair dealing for the purpose of private study, research, criticism or review, as permitted under the Copyright Act, no part may be reproduced by any process without written permission. Inquiries should be addressed to the publisher.

National Library of Australia
Cataloguing-in-Publication entry:

   Murray, David R. (David Ronald), 1943– .
   Seeds of concern: the genetic manipulation of plants.

   Includes index.
   ISBN 0 86840 460 8. (UNSW Press)
     ISBN 0 85199 725 2 (CABI)

   1. Transgenic plants. 2. Plant genetic engineering.
   I. Title.

   631.5233

A catalogue record for this book is available from the British Library.
A catalogue record for this book is available from the Library of Congress, Washington, DC, USA.

Cover design Di Quick
Printer BPA

CONTENTS

Preface	7
Acknowledgments	9
Abbreviations and acronyms	11
1 Introduction: Cells, genes and chromosomes	13
2 How genetically modified plants are produced	31
3 The hazards of herbicide-resistant plants	44
4 Setting priorities for plant improvement	59
5 Proposals with nutritional, medical or utilitarian goals	74
6 Environmental and health impacts of genetically modified plants	85
7 Intellectual property issues	99
8 Impacts of genetically modified plants in the Third World	115
9 Loose ends	129
Useful addresses	138
Glossary	142
Further reading	148
Index	150

PREFACE

Several popular books about the implications of gene technology have appeared in recent years, but none has dealt comprehensively with genetically modified plants. Most of the adverse publicity about genetically modified organisms concerns plants. How much of the controversy is justified?

This book arose from my concern to update topics canvassed in *Advanced Methods in Plant Breeding and Biotechnology* (1991), and to convey this basic information more readily to interested members of the public. I have described what has been attempted with recombinant nucleic acid technology, explained what is wrong with what has been done so far, and indicated how things could have been done differently. There are some worthwhile objectives that might still be accomplished, and these too are discussed. What I have suggested is that every proposed release of a genetically modified plant should be judged on its merits, rather than being approved automatically by 'rubber stamp' committees, or opposed automatically for no sound reason.

Breaking down the mythology and misconceptions fostered by some of the biggest players is an important part of this book. Some people are concerned about the safety of the procedures used by this industry, and the industry's encouragement of ecologically unsustainable agricultural practices. Many people are also concerned about corporate monopoly of genetic resources through overly restrictive laws concerning intellectual property and world trade agreements. The multinational companies that dominate trade in seeds perceive ownership of plant genes as a way to increase profits. This aspect of globalisation intrudes on the self-sufficiency of farmers in many countries and has disruptive social consequences. Such exploitation can no longer be justified.

If you are concerned about the possible impacts of genetically modified plants on genetic diversity, the environment, human health, or human society, then here is a balanced source of information. Uncritical proponents of genetically modified organisms often express the wish for a better informed public debate. This book is a contribution to that objective.

David R Murray

ACKNOWLEDGMENTS

Many people have contributed in various ways to the writing of this book. For helpful discussions and encouragement, I thank Peter Abell, the late Senator Robert Bell, Dr Judy Carman, Daniel Deighton, Dr Heather Dietrich, Dr Margaret Dwyer, Jude Fanton, Michel Fanton, Rayyar Farhat, Ieva Gay, Bill Hankin, Professor Stuart Hill, Leila Huebner, Sue McGregor, Dr Judyth McLeod, Gayle Murray, Dr Ray Ritchie, Dr Roger Spencer, Andrew Storrie, the late Fay Sutton, and Dr Claudia Tipping. For providing copies of articles or lending or donating books, I thank Dr Keith Brown, Leesa Daniels, Dr Margaret Dwyer, Ieva Gay, Bill Hankin, Professor Stuart Hill, Dr Judy Messer, Lyndall McCormack, Dr Judyth McLeod, Dr Helene Martin, Dr Matthew Morell, Dr Frank Peters, Bob Phelps, Dr Alan Richardson, Andrew Storrie, Arnold Ward and Marion (Mazza) Welham.

I am particularly grateful to Dr Allan Green, Dr TJV Higgins, Dr Danny Llewellyn, Dr Matthew Morell, Rachael Mitchell, Dr Alan Richardson and Dr Iain Wilson for discussing their projects with me during a visit to CSIRO Plant Industry in May 2001, and for allowing me to take photographs. I also thank Peter Abell for hosting a visit by members of the Australian Plants Society to the University of Sydney Plant Breeding Centre at Cobbitty, NSW, and for later checking the labelling of my photographs.

For hospitality, I thank Jude and Michel Fanton (Byron Bay) and Bill Hankin (Adelaide). I also thank the Australian Plants Society (NSW) for supporting my attendance at an Australian Cultivar Registration Authority meeting at Adelaide Botanical Gardens (2000), and Heritage Seed Curators Australia for their support of an earlier visit to Adelaide on the occasion of the 11th Australian Plant Breeding Conference (1999). It was immediately after that conference that I submitted the proposal for this book.

A number of scientists provided answers to queries and copies of papers. I am grateful to them, and to the following for permission to reproduce photographs or other illustrations: Dr Marc De Block (Figure 2.1), Daniel Deighton (Plate 18), Jude and Michel Fanton (Plates 24–29), Dr Ian Heap (Figure 3.1), and Dr Claudia Tipping (Figure 1.3). Unless otherwise acknowledged, the photographs are my own.

Finally, I would like to express my thanks to John Elliot of UNSW Press for supporting this book at every stage of its development.

ABBREVIATIONS AND ACRONYMS

ACRA	Australian Cultivar Registration Authority
ANZFA	Australia and New Zealand Food Authority
Bt	*Bacillus thuringiensis*
CaMV	cauliflower mosaic virus
CGIAR	Consultative Group on International Agricultural Research
CIMMYT	Centro Internacional de Mejoramiento de Maiz y Trigo, Mexico
CIP	International Potato Centre, Lima
CSIRO	Commonwealth Scientific and Industrial Research Organisation
2,4-D	2,4-dichlorophenoxyacetic acid
DDT	dichloro diphenyl trichloroethane
DNA	deoxyribonucleic acid (or deoxyribose nucleic acid)
EU	European Union
F_1	first filial generation
FAO	Food and Agriculture Organisation (United Nations)
FSANZ	Food Standards Australia and New Zealand
GMAC	Genetic Manipulation Advisory Committee
GMO	genetically modified organism
GTCCC	Gene Technology Community Consultative Committee
GTEC	Gene Technology Ethics Committee
GTTAC	Gene Technology Technical Advisory Committee
GUS	ß-glucuronidase
HSCA	Heritage Seed Curators Australia
ICRISAT	International Crops Research Institute for the Semi-Arid Tropics
IOGTR	Interim Office of the Gene Technology Regulator
IRRI	International Rice Research Institute (The Philippines)
MHR	Member of the House of Representatives

NASAA	National Association for Sustainable Agriculture Australia
PBR	Plant Breeders Right (or Rights)
PPO	polyphenol oxidase
PVR	Plant Variety Right (or Rights)
RAFI	Rural Advancement Foundation International
RHS	Royal Horticultural Society
RNA	ribonucleic acid (or ribose nucleic acid)
SD	standard deviation
SSN	Seed Savers' Network
2,4,5-T	2,4,5-trichlorophenoxyacetic acid
TRIPS	Trade Related Intellectual Property Rights
UNDP	United Nations Development Program
UNESCO	United Nations Educational, Scientific and Cultural Organization
UPOV	International Union for the Protection of New Varieties of Plants
USDA	United States Department of Agriculture
VACVINA	Vietnamese Community Action Programme Against Hunger, Malnutrition and Environmental Degradation

1
INTRODUCTION: CELLS, GENES AND CHROMOSOMES

> Such is life.
>
> Ned Kelly

CELLS AND THEIR COMPONENTS

News items concerning cells and DNA are broadcast almost every day. We take for granted the knowledge that complex living organisms consist of cells and specialised tissues, which grow and change at different stages of development. But this insight is comparatively recent. Using simple light microscopes, biologists began to establish the multicellular nature of complex organisms just over 300 years ago. Advances in optics in the Netherlands early in the 17th century allowed both telescopes and microscopes to be improved. English, Dutch and Italian scientists first took advantage of these microscopes to delve into the structure of living organisms.

Why do we use the word 'cell'? The English scientist Robert Hooke (1635–1703) observed spaces in thin sections of cork tissue and called them 'cells' in his publication *Micrographia* in 1665.[1] The sense in which he used this term is the same as for our gaol cell, as his cork cells were simply chambers devoid of contents. What he described was a matrix of external cell walls, typical of most plant tissues. Marcello Malpighi (1628–1694) and Nehemiah Grew (1641–1712) were the first to describe plant tissues in terms of their constituent cells, both publishing their observations in 1671.[2] Subsequently Anton van Leewenhoek (1632–1723) is credited with the first observations of human sperm cells and bacteria in 1674.[2] Nehemiah Grew published a further treatise on plant anatomy in 1682, and was one of the first to study the varied shapes and sizes of pollen grains.

Details of cell structure have emerged progressively since the beginning of the 19th century. Although a general 'cell theory' is often attributed to Matthias Schleiden (1804–1881) and Theodor Schwann (1810–1882) because of their pronouncements in 1839, earlier writers had also drawn attention to the cellular basis of tissues, for example, the zoologists Lorenz Oken in 1805,[3] and Jean-Baptiste de Monet de Lamarck in 1809.[4] The botanist Robert Brown (1773–1858), who accompanied Matthew Flinders in the circumnavigation of Australia between 1801 and 1803, identified the nucleus in 1831.[4,5] Furthermore, he reported the occurrence of a nucleus as a constant feature of almost every cell. The nucleus is surrounded by cytoplasm, and the movement of cytoplasm around a living cell was evidently first recorded by Wilhelm Hofmeister in 1867.[6] The dynamic nature of the living cell is often overlooked as we study micrographs or line diagrams, which can only represent 'snapshots' of a thin slice of the cell at a given instant.

Originally the term 'protoplasm' was applied to everything inside the cell wall. Then in 1882[7] 'cytoplasm' was applied to everything in a plant cell except the nucleus and the vacuole, a central compartment containing sap and sometimes pigments. Since the advent of electron

Figure 1.1
An idealised diagram of a plant cell

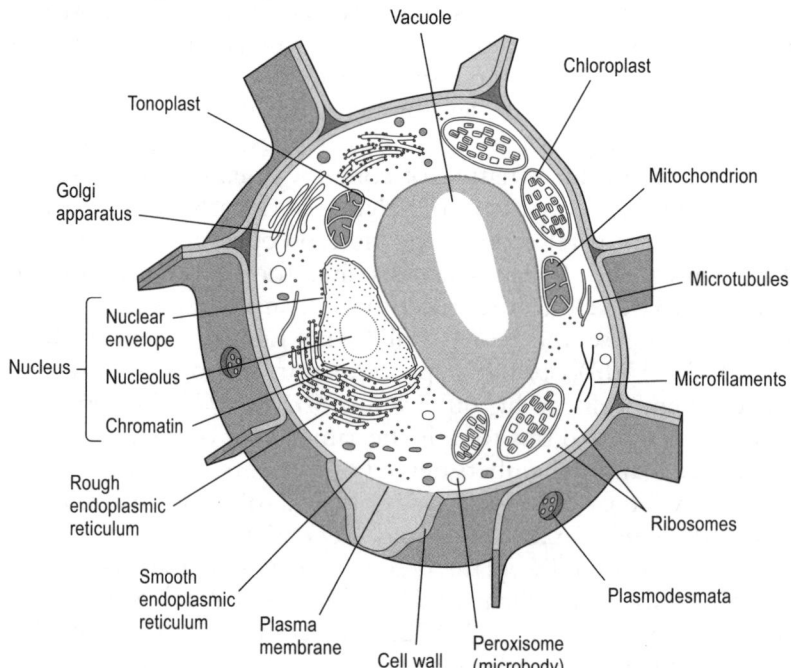

microscopes in the middle of the 20th century, more and more of the cytoplasm has been found to possess structure. We need to take account of this detail before considering the ways transgenic plants are produced (Chapter 2).

The larger cellular inclusions are the membrane-bounded organelles (Table 1.1; Figure 1.1). The various parts of a cell are adapted to performing different functions. Just as organs of the whole plant are adapted primarily for photosynthesis, storage, nutrient uptake or reproduction, so each kind of organelle carries out specific functions within a cell, and their membranous barriers provide control of transport and metabolism (Table 1.1). Complex cells of this kind are termed 'eukaryotic' ('with a true nucleus') to distinguish them from 'prokaryotic' cells with a simple nucleus (nucleoid) that lacks a bounding membrane.

Table 1.1
The main subcellular components of plant cells

Organelle or structure	Major functions
Nucleus	Inheritance; control of gene expression, cell differentiation and metabolic activities
Vacuole	Control of turgor (cell rigidity); storage of minerals, pigments, proteins, tannins, and some crystalline substances; breakdown of reserves following seed germination
Microbodies	Oxygen assimilation; amino acid metabolism; conversion of fatty acids to sugars
Plastids	Photosynthesis (chloroplasts); attraction (chromoplasts in flowers and fruits); starch storage (amyloplasts and chloroplasts)
Mitochondria	Respiration, energy conversion and biosynthesis
Golgi bodies (dictyosomes)	Processing and transport of complex macromolecules to destinations inside or outside the cell
Spherosomes	Storage of oils, especially in seed tissues
Smooth endoplasmic reticulum	An internal membrane system allowing further compartmentation (separation) of metabolic pathways
Rough endoplasmic reticulum	Ribosomes attached to smooth endoplasmic reticulum
Ribosomes	The sites of polypeptide synthesis
Microtubules	Contractile movements (cytoplasmic streaming)

The nucleus has remained the nucleus, but the term 'cytoplasm' now presents difficulties. Light microscopists still tend to call the transparent parts of the cell the cytoplasm, but strictly the soluble phase of the cytoplasm should now be called the 'cytosol'. The term 'cytoplasm' is historically important, and retained in the phenomenon of cytoplasmic inheritance encountered by plant breeders (see below).

The conclusion that living cells arise only by division from pre-existing cells is very important, but it took almost the whole of the 19th century to become generally accepted. Logic and intuition were not sufficient. A crucial step came in 1861, when Louis Pasteur (1822–1895) showed that the breakdown of meat broth in flasks with S-shaped necks depended on the presence of live bacteria.[1,4] In flasks sterilised by boiling, no breakdown occurred unless the S-shaped neck was snapped off, readmitting bacterial spores from the air (*'les germes qui flottent dans l'air'*).[8] So the clear meat broth did not spontaneously generate the organisms responsible for its breakdown.

DNA AND THE GENETIC CODE

How can something as small as the nucleus of a cell control the metabolic activity and properties of that cell, and ultimately the properties of a complex, multicellular organism? The answer lies at the molecular level, below the resolution of most microscopes. By chemical analysis, the nucleus is known to contain deoxyribonucleic acid (DNA) and structural proteins called histones. On hydrolysis, the DNA component yields a sugar (deoxyribose), inorganic phosphate and four distinct nitrogenous bases: the purines, adenine and guanine; and the pyrimidines, thymine and cytosine. How can this simple analytical result account for the ability of the nucleus to regulate complex activities and provide for inheritance of an organism's 'blueprint' from generation to generation?

In 1953 Linus Pauling suggested a helical structure for DNA, similar to the alpha helix he had successfully proposed for polypeptides. He placed a repeating deoxyribose-phosphate backbone in the centre, with the nitrogenous bases on the outside, and suggested that three such strands were woven together.[9] But many features of this model were unsatisfactory; it lacked the ability to explain or predict.

In the same year, James Watson and Francis Crick[10] proposed a model comprising a double helix. Each strand of DNA in this double helix consisted of a long polymer that had repeating deoxyribose and phosphate groups, but with the attached nitrogenous bases projecting to the interior at regular intervals, so that ten base pairs occurred in a 360° sweep of the double helix. They proposed that the bases of one strand form complementary pairs with the bases of the opposite strand, so that adenine always pairs with thymine

(A–T), and guanine always pairs with cytosine (G–C). In this way the structure is stabilised by the greatest possible number of hydrogen bonds.

This model was able to explain how DNA could reproduce itself.[9,11] As the helices separate, each strand acts as a template for the assembly of complementary nucleotide precursors, positioning these correctly before polymerisation takes place (Figure 1.2). Because both original strands are conserved when their complements are newly synthesised, the mechanism of DNA replication was termed 'semiconservative'.

Figure 1.2
A 'semi-conservative' model to explain DNA replication, adapted from J. D.

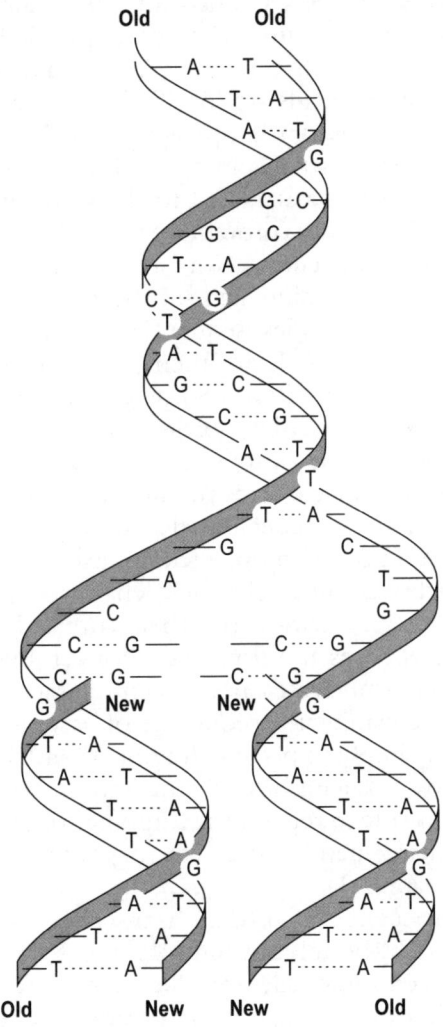

Watson's book The Double Helix[9]

The Watson–Crick model laid the foundation for breaking the genetic code. For a sequence of bases in one strand of DNA to specify the sequence of amino acids in a polypeptide, various kinds of ribonucleic acid (RNA) are first synthesised from template DNA: messenger RNA (mRNA), which moves between the DNA and the ribosomes where polypeptides are assembled; ribosomal RNA (rRNA), which is a structural part of each ribosome; and numerous forms of transfer RNA (tRNA), which ferry individual amino acids to their correct positions. The synthesis of RNA from a DNA template is called 'transcription'.

The pyrimidine base uracil (U) occurs in RNA instead of thymine. Like thymine, uracil is complementary to adenine. To cut a long story short, the 'codons' of mRNA consist of sets of three bases. There are 64 possible sets. Only two amino acids are specified by a single codon: tryptophan (UGG) and methionine (AUG). The other 18 common protein-forming amino acids are specified by up to six codons each. In addition, three codons are stop signals: UAA, UAG and UGA. Transfer RNA molecules have an anticodon region of three complementary bases that can be attracted to the appropriate codon regions of mRNA. Special enzymes (protein catalysts) join amino acids to their appropriate tRNA molecules. Not surprisingly, these enzymes are highly specific for their amino acid substrates.[12] At the ribosomes, the appropriate tRNA molecules sequentially pair with the codons in mRNA, and the amino acids are then joined to form a polypeptide; this process is called 'translation'. So by governing the base sequences in mRNA and tRNA molecules, portions of the DNA ultimately determine the sequences of the various amino acids in polypeptides.

The synthesis of nucleic acids requires enzymes called polymerases to make the initial joins between nucleotides. In addition, nucleic acid molecules undergo processing by nicking, excision, and rejoining (ligation). Endonucleases cut nucleic acid chains at specific points. They have different site-specificities. In other words, they recognise a particular sequence of bases and usually do not act unless this sequence is present. Many endonucleases have been characterised and they can now be used to determine the sequences of bases in DNA from diverse sources,[13,14,15] or simply to provide fragments of DNA for comparative studies (see below). Ligases are enzymes that rejoin breaks in nucleic acids. Besides this role in repair or recombination, they are now important for introducing gene constructs to genomes being deliberately transformed (Chapter 2).

Extensive processing of mRNA 'transcripts' occurs in eukaryotes (most organisms), although not in prokaryotes such as bacteria. Some parts (introns) are removed, and the remaining parts (exons) are rejoined.[13,16] Minor variations in the positions where excisions begin or end can give rise to 'isoforms' of proteins that might have different locations within a cell, or subtle differences in properties that make them more suitable for specific tasks in specialised tissues.[17] After translation, usually a number of times, mRNAs are broken down and their components re-used. Introns are constantly being re-used. The other forms of RNA are more stable, but all are ultimately broken down by specific enzymes and recycled.

GENES AND GENOMES

At a simple level, a gene is 'a discrete unit of inheritance represented by a length of DNA located in a chromosome'.[18] Usually a gene specifies an enzyme, a structural protein, or an RNA transcript of some kind. A gene is confined to one strand; the complementary strand does not code for anything.[16] A 'genome' is the complete collection of genes and non-coding DNA sequences belonging to a given organism. The term can be qualified according to whether one is considering the nuclear genome, an organelle genome or the whole genome.

The chromosomes contained within the nucleus contain most of an organism's heritable material, but not all of it. Some very important genes are located in the circular DNA of the chloroplasts or related plastids and the mitochondria (Figure 1.3). Why should these organelles contain DNA, and why does its organisation resemble the circular chromosome of a bacterium? How do we know that these organelles do not just acquire some bacterial DNA as contamination whenever they are isolated from cell and tissue debris?

Figure 1.3
Part of a leaf cell showing nucleus (n), vacuole (v), mitochondrion (m) and chloroplasts (c) containing starch granules (s). Electron micrograph courtesy of

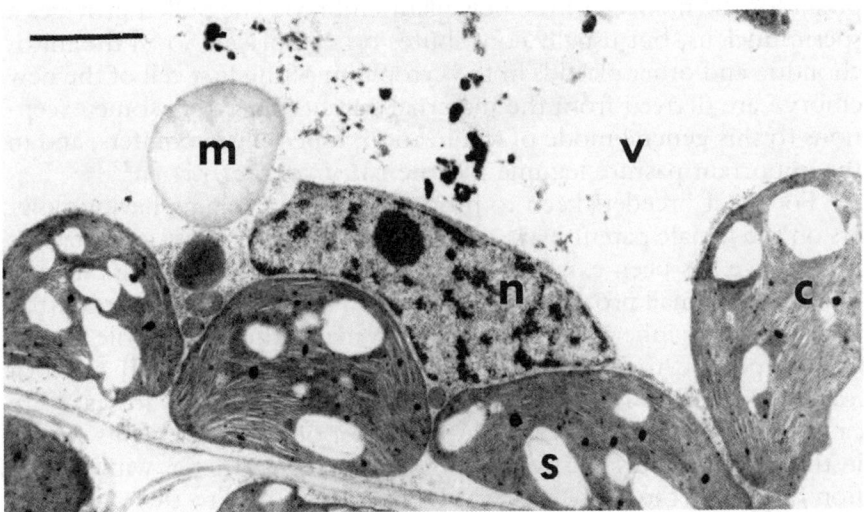

Dr Claudia Tipping.

Organelle DNA is not an artefact of isolation. Evidence gathered over the past 50 years is fully consistent with the idea that ancestral eukaryotic cells first acquired proto-organelles by engulfing other prokaryotic cells, then failing to digest them.[19] The trapped 'endosymbionts' have become the microbodies, mitochondria, chloroplasts and related plastids of modern plant cells.

Chloroplasts and mitochondria retain only a small portion of their original genetic information — most has been redistributed to the nuclear chromosomes. The synthesis of chloroplast and mitochondrial proteins now involves a close co-ordination between nuclear and plastid genomes. In no sense are these organelles autonomous or even 'semi-autonomous', a common assumption in the 1960s. An example of this co-ordination involves the enzyme chiefly responsible for fixing carbon dioxide during photosynthesis, ribulose *bis*phosphate carboxylase. This enzyme is located inside the chloroplasts, but only its large subunit is manufactured there; its small subunit is made at ribosomes in the cytoplasm. The two kinds of subunit are assembled into functional proteins inside the chloroplasts. The large subunit is coded in the chloroplast genome and the small subunit is coded in the nucleus.

The chloroplast genome is better understood than the mitochondrial, and consists of 120 to 160 kilobase pairs, containing approximately 113 to 127 genes.[20,21] A representative plant mitochondrial genome contains only 90 genes.[22] Over millions of years, the coding pattern in plastids has shifted away from sequences typical of bacteria to sequences more like those of the plant nuclear genome.

Plastid and mitochondrial genomes are responsible for so-called 'cytoplasmic' or maternal inheritance, which occurs in most higher plants. At fertilization, the egg cell provides all or most of the cytoplasm for the first cell of the new plant embryo. The pollen provides a sperm nucleus, but usually contributes no cytoplasm. So all the mitochondria and other plastids in the cytoplasm of the first cell of the new embryo are derived from the maternal parent. There are some exceptions to this general mode of fertilization, especially in conifers, and in the important pasture legume lucerne (*Medicago sativa*).[20,22,23]

For plant breeders keen to produce hybrids easily by having flowers on the female parent plant endowed with male sterility, cytoplasmic inheritance has been extremely important. One form of male infertility involves a small protein in the mitochondrion, and so is transmitted by cytoplasmic inheritance. However, in maize the male-sterile condition ('type T' cytoplasm) coexists with susceptibility to Southern corn leaf blight (*Helminthosporium maydis*). Massive crop losses were caused by this disease in 1970,[24] when most of the maize plants grown in the United States had type T cytoplasm. New varieties with a deletion of part of the mitochondrial DNA are resistant to the toxin produced by this fungus, and remain male fertile.[25]

CHROMOSOMES

In August 2000, eight contestants on 'Who Wants to be a Millionaire'[26] were asked about the distinguishing chromosome responsible for male–female differences in humans. Only two contestants correctly chose the Y chromosome as their answer from four possibilities. In other words, 75 per cent of respondents were incorrect. This is a small sample, but one biased in favour of people who think they have a good general knowledge. Such a result extrapolated to the whole population would indicate a general level of ignorance about genetics that is quite deplorable. Small wonder that our parliamentarians are beguiled by the simplistic assurances of lobbyists who are keen to place their commerce above the community's best interests.

Chromosomes were first visualised in the late 1880s, when German microscopists developed staining procedures that revealed their structure. Chromosomes are the packaging units of the nuclear genome. They are supercoiled nucleic acid–protein complexes, and become visible in this fashion just prior to and during cell division. Their sizes vary enormously, as does the number typical of a given species, called the 'karyotype' (Table 1.2).

Table 1.2
Chromosome numbers of some important food plants[30]

Species	Karyotype (and genome)	Haploid number
Dicotyledons		
faba bean (Vicia faba)	2n = 12	6
pea (Pisum sativum)	2n = 14	7
chickpea (Cicer arietinum)	2n = 16	8
onion (Allium cepa)	2n = 16	8
carrot (Daucus carota)	2n = 18	9
kale (Brassica oleracea)	2n = 18 (CC)	9
turnip (Brassica campestris)	2n = 20 (AA)	10
swedes, rapes (Brassica napus)	2n = 38 (AA, CC)	19
common bean (Phaseolus vulgaris)	2n = 22	11
cowpea (Vigna unguiculata)	2n = 22	11
tomato (Lycopersicon esculentum)	2n = 24	12
capsicum (Capsicum annuum)	2n = 24	12
soybean (Glycine max)	2n = 40	20
Monocotyledons		
barley (Hordeum vulgare)	2n = 14	7
rye (Secale cereale)	2n = 14	7
goat grass (Triticum tauschii)	2n = 14 (DD)	7
emmer wheat (Triticum turgidum)	2n = 28 (AA, BB)	14
bread wheat (Triticum aestivum)	2n = 42 (AA, BB, DD)	21
maize (Zea mays)	2n = 20	10
sorghum (Sorghum bicolor)	2n = 20	10
rice (Oryza sativa)	2n = 24 (AA)	12

Having two sets of homologous chromosomes is the normal condition for vegetative cells throughout a plant. The number of chromosomes in a set is called the haploid number, n, and double this, the diploid number, is 2n. Egg cells and sperm cells are reduced to the haploid number by meiosis (reduction division) during their formation, then the diploid condition is recovered on fusion of a sperm cell with an egg cell. Polyploidy, as in *Brassica napus* or bread wheat (Table 1.2), can occur when natural crosses between different species are successful and stable. Swede turnips and rapes resulted from the spontaneous crossing of kale with turnip, possibly on many occasions. Bread wheat arose from a diploid parent (*Triticum tauschii*, also called *Aegilops squarossa*), which crossed with a cultivated tetraploid type similar to emmer or durum wheat. This has been confirmed by deliberately repeating the cross, opening the way for the introduction of genes for disease or pest resistance from *Triticum tauschii* to bread wheat.[28] In cases like these, where the nuclear genomes from specific sources have been identified, they are distinguished by capital letter (Table 1.2).

CONVENTIONAL PLANT BREEDING

Although selection has been going on for thousands of years, the deliberate breeding of plants is a relatively young science, dating from about 1780. Thomas Andrew Knight (1759–1838), an Englishman, developed the two-step procedure of hybridisation and selection long before there were genetic explanations of why this technique should be so successful. He prevented uncontrolled pollination, whether from selfing or external sources, and used known pollen donors. He was able to generate many more variants than usual, and from these selected plants with the most desirable combinations of characters.

Peas were normally round-seeded, starchy and bland, harvested at maturity for storage and later consumption as soup or pease pudding. Knight developed sweeter peas with wrinkled seeds from 1787 onwards. His new peas came to be highly regarded, and over the next half-century he revolutionised green peas as a vegetable. Through his good friend Sir Joseph Banks, one of Knight's new varieties was transmitted to Australia with Philip Gidley King when he returned as Governor of New South Wales in 1800. This is the Tall Marrowfat that King records in his correspondence with Lord Hobart in 1803.[29] Knight also bred many new kinds of fruit tree, and several notable strawberries, such as the Downton (1817) and the Elton (1828). The latter also made its way to New South Wales.[30]

Knight forced his fruit tree seedlings to flower sooner by grafting them onto well-established rootstocks, saving many years in the process. His *modus operandi* became very well known, and was widely adopted in the United States following the publication of his book *Treatise on the Culture of the Apple and Pear and on the Manufacture of Cider and Perry* in 1806. This book ran to at least a third edition, which was published in 1808. Extracts were published weekly in a periodical called *The Rural Visiter*, begun by David Allinson at Burlington, New Jersey, in July 1810. There is no doubt that later American plant breeders such as Charles Hovey and Luther Burbank drew their inspiration and most productive techniques from Knight's example, as did a multitude of English pea breeders.[31]

For a long time the empirical plant breeders went their own way, oblivious to the scientists who were studying the processes of pollen grain formation, fertilization and inheritance. The discoveries of Wilhelm Hofmeister (1824–1877) and Gregor Mendel (1822–1884) had profound implications for plant breeders, but little notice was taken of their insights until after 1900.

Using a microscope and cutting thin slices of still-living (unfixed) plant material, Wilhelm Hofmeister observed the details of pollen grain germination, pollen tube growth and fertilization in representatives of 19 families of flowering plants, and published these results in 1849.[6] He extended earlier observations on orchids by Amici and von Mohl, and concluded that a new embryo forms when a sperm cell coming through the pollen tube fuses with an egg cell inside the ovule. Hofmeister was a self-taught German with no formal tertiary education. He was able to publish his observations through his father's printery, which normally produced musical scores.[6]

Then the Augustinian monk Gregor (Johann) Mendel selected the pea plant as the vehicle of his personal demonstration of the validity of Hofmeister's conclusions — with amazing results. Mendel studied peas at the monastery of St Thomas in Brno, Moravia (then Brünn, under Austrian government). It is well known that he published his findings in an obscure local journal of natural history in 1866 — and they sat on the library shelf in various institutions until rediscovered 34 years later. Mendel's paper, *Experiments in Plant Hybridization*, was not published in English until translated by William Bateson for the Royal Horticultural Society.[32] Only recently, however, has light been shed on Mendel's motivation for doing his research.

Far from being the objective, dispassionate investigator isolated in his monastery garden, Mendel was highly motivated. He was furious at being failed in his Botany examination at the University of Vienna in 1856 by the ultra-conservative Professor Fenzyl, who had refused to accept Hofmeister's general conclusion about fusion of sperm and egg cells. Fenzyl still believed that the new plant embryo was an outgrowth of the pollen tube, an earlier but inaccurate conclusion drawn by the influential Professor Schleiden.[6] This whole episode is redolent of the conflicting Greek views about human reproduction — Hippocrates (460–375 BC) holding that a foetus arose from the union of male and female 'seeds', but Aristotle (384–322 BC) regarding the female only as a vessel or receptacle, with the foetus being derived from the sperm. Hippocrates was right — and so was Hofmeister.

Gregor Mendel went to his monastery insulted and determined to prove a point.

That is why he was so sure in his assumption that equal contributions to inheritance are made by both the female and the pollen parents. In turn, this assumption allowed him to discern the concept of dominance and recessivity when the F_1 hybrids of his crosses totally submerged some characters in favour of others, only for them to reappear in subsequent offspring derived once again by self-fertilization.

It is nonsense to suggest, as the statistician R. A. Fisher has,[33] that Mendel's results are 'too good to be true' or that he could not really tell the difference between yellow and green embryos. Mendel's results are entirely in keeping with the careful way he went about his study. He made preliminary observations over two years. Out of 34 pea varieties obtained from a number of seedsmen, he then selected only 22 'true-breeding' kinds to be the parents in his hybrid crosses. He showed that these 22 kinds remained true-breeding over the entire eight-year period of his experiments. He also had to contend with a more complicated taxonomy than we do. Some of his varieties were known by different species names, such as *Pisum saccharatum* for peas with a 'snow pea' pod, or *Pisum umbellatum* for those whose flowers were crowded at the top of the plant.

Disregarding the taxonomy, Mendel chose characteristics that were readily distinguished from one another (Table 1.3). His conclusions about which were dominant, and which recessive, were correct, and his shorthand symbolism for inherited factors (now called genes) is accepted to this day.

Table 1.3
The original 'Mendelian' characters of pea plants

Dominant characteristic	Corresponding recessive condition
Tall plants with long internodes	Dwarf plants with short internodes
Flowers from axillary shoots	Flowers terminally clustered
Flowers violet and mauve/purple[a]	Flowers white
Seed-coats opaque and pigmented[a]	Seed-coats not strongly pigmented
Seed shape round, or slightly dented	Seeds strongly wrinkled[b]
Pods uniformly inflated	Pod walls constricted around seeds
Pods green	Pods yellow[c]
Mature embryo turns yellow	Mature embryo remains green

[a]These characters were firmly correlated in Mendel's crosses but are now known to involve more than just a single pigment gene.
[b]This difference is now known to involve complex changes in starch and protein composition, as well as 'concertina' cell walls.
[c]In common beans the same condition gives rise to wax pods or butter beans.

As an example, consider a cross between two pure-breeding peas, one with yellow embryos and the other with green. The F_1 hybrid produces peas with only yellow embryos. But in the next (F_2) generation obtained by self-fertilization, pea seeds with yellow or green embryos are produced in a ratio of 3:1 respectively. Mendel's actual numbers from 258 F_1 plants were 6022 seeds with yellow embryos and 2001 with green, a ratio of 3.01 to 1.

Representing the dominant factor for yellow embryos as Y, and the recessive factor for green as y, these results could be explained if the original parents had factors YY and yy, respectively, and their F_1 hybrid had Yy, with one factor donated by each parent. At flowering, the F_1 hybrid would be producing two types of egg cell (Y or y) in equal proportions, and two kinds of pollen grain, Y or y, again in equal proportions. These alternative inherited factors affecting a character are now called 'alleles'. By chance, the four possible combinations of egg cell and sperm cell should also occur in equal proportions (Table 1.4). Thus all the F_2 peas with green embryos must be true-breeding (yy). However, only one-third of the seeds with yellow embryos would be true-breeding YY like the original parent; two-thirds would be Yy like the F_1.

Table 1.4
Combinations of egg and sperm cells giving rise to yellow and green embryos in garden pea

		Sperm cell genotypes (50% each)	
		Y	y
Egg cell genotypes (50% each)	Y	YY	Yy
	y	Yy	yy

Mendel also swapped the parents around, making similar crosses with first one, then the other, as pollen donor. He showed that this makes no difference to the outcome. This was a crucial observation in support of inherited factors being transmitted via the gametes. He also counted results of some crosses looking at two pairs of characters at once, for example embryo colour and round versus wrinkled seed shape. From such results he derived the principle of independent assortment of inherited factors during the formation of pollen grains and egg cells, that is, whether an embryo is green or yellow has no effect on whether it is round or wrinkled, and vice versa. He was fortunate to have avoided the complication of linkage, which reflects how close together genes might be within a chromosome, and maternal inheritance (discussed earlier). Gregor Mendel provided a marvellous beginning. His scientific career was brief, but his contribution to our understanding of genetics was immense.

GENE MAPPING AND GENOMICS

Mapping genes to positions on chromosomes had been going on for decades before techniques for gene sequencing became available. Many genes with alternative alleles have provided invaluable markers for developing linkage maps.[15] The old-fashioned methods involved crossing varieties with known alleles and measuring the extent to which their inheritance differed from the proportions expected from random assortment of alternative alleles to egg or sperm cells. In other words, deviations from Mendel's principle of independent assortment were measured arithmetically and applied arbitrarily to construct linkage 'distances'. These distances do not exactly reflect numbers of bases along a DNA molecule.

Another kind of observation has also been made over many years. This is the measurement of the amount of DNA that plant cells characteristically possess. The 'C-value' is the amount of DNA belonging to a haploid nucleus, expressed in picograms (10^{-12} g). This is an indicator of the size of the nuclear genome. It has become clear that the size of the genome has often increased as flowering plants (Angiosperms) evolved.[34,35] However, the magnitude of the differences between species cannot be explained simply by multiple extra copies of the genes held in common by all higher plant species. Major differences result from variable amounts of highly-repeated 'spacer' sequences — the material once dubbed 'junk DNA'. In some species, this non-coding fraction accounts for most of the DNA. But now it is clear that variation in non-coding repeated DNA sequences 'may also define species differences and drive evolution'.[35]

The impetus to map and sequence genes gathered pace in the early 1990s, with co-operative efforts launched to develop complete maps for species such as rice, maize, tomato, pea, and a small weed called *Arabidopsis thaliana*, which has a rapid generation time and a relatively small genome.[15,36] Some of these genomes have now been completely sequenced.[37] Recently a consortium has formed with the aim of elucidating the genome of banana and making the results publicly available.[38] Genomics has become equated with determining the complete base sequence of the nuclear genome of any given organism. The human genome was sequenced by many teams over about 10 years, culminating in announcements made prematurely on 26 June 2000,[39] then repeated in February 2001. But we do not need to know the complete sequence of a genome to gain useful insights.

Comparing the details of base sequences of common genes permits one approach to determining plant relationships. Phylogeny seeks to discover relationships in terms of descent from common ancestors, and one or a few genes can be studied rather than the whole genome. The basic assumption of this approach is that fewest sequence differences in any particular gene are shown by the species (or varieties within a species) that are most closely related. Because different genes have acquired random alterations at different average rates,[21] it is a good idea to study more than a single gene. Hypothetical pedigrees can be constructed so that any set of species can be arranged in the most economic (parsimonious) way possible.

An excellent illustration of the effectiveness of this approach is a study using fragment patterns of chloroplast DNA to elucidate relationships among tomatoes, potatoes and allied species in the family Solanaceae.[40] Different fragment patterns result from treatment of DNA preparations with a range of endonucleases with distinct specificities. Changes to bases through mutation are reflected in the resulting patterns. The derivation of tomato (*Lycopersicon esculentum*) from a species of *Solanum* has been worked out so clearly that the question of changing the genus name back to *Solanum* has become an issue.[40,41]

Conversely, some studies waste opportunities to gain valuable insights about relationships, and draw incorrect conclusions. One recent study of *Acacia* included only four Australian species with phyllodes (flattened leaf stalks) in a sample of 68 species world-wide,[42] despite the fact that most species of *Acacia* are Australian (more than 1000 species).[43]

Having the technical ability to obtain molecular data does not automatically endow researchers with the skills needed for experimental design, logical deduction and correct interpretation of results. Botanists today need to comprehend all other kinds of information about plants before attempting to interpret DNA sequence data, and then they need to proceed cautiously.[44]

THE JURASSIC PARK SYNDROME

The idea that extinct organisms might be brought back to life from preserved DNA has caught the public imagination. Steven Spielberg's films applied this scenario to dinosaurs. In the United States, companies exist that will freeze the bodies of dead pets against the day when it might be possible to clone from some of the preserved cells. The pets' owners will not live long enough for this to happen, but other companies will freeze them in the hope of eventual resuscitation. Cryogenics is booming. And scientists who should know better are proposing to resurrect extinct organisms from tiny amounts of preserved DNA.

Resurrecting the Tasmanian tiger (*Thylacinus cynocephalus*) from an animal pickled in a museum jar since the mid-19th century has recently been the subject of a well-publicised proposal[45] that falls far short of feasibility. For a start, the condition of the DNA is problematic, given that alcohol is not as good a fixing agent as amber. Suggesting 'five to twenty-five years' as a time frame for such an endeavour is hopelessly optimistic (Michael Archer, quoted by Rebecca Lang).[45] Continuing this project would be extremely wasteful of limited resources.

Nevertheless, phylogenetic studies of the kind discussed above can usefully be extended back to include species from almost 100 million years ago. This becomes possible when organisms have been preserved in amber, which forms after they have become trapped in sticky plant gums. If the amber hardens quickly enough, it protects the enclosed organisms against breakdown by aerobic bacteria and maintains the structure of their DNA. New Jersey amber, dating from 94 to 90 million years ago, contains oak-like flowers in an excellent state of preservation, as well as many insects.[46] The DNA coding for rRNA from some of these preserved insects has been analysed, and the relationships of the preserved insects to modern species confirmed.[46] The possibility exists for similar studies of ancient Angiosperms, and these would be invaluable for testing proposed relationships.

Finally, it needs to be made clear that a knowledge of DNA sequences gives no information about the organisation of that DNA at the level of individual chromosomes or organelles. And without viable cells, DNA sequence information comes to a dead end.

Make no mistake: extinction is forever.

REFERENCES

1 Pledge, HT (1959) Evolution and the microscope. In HT Pledge *Science Since 1500*. Harper, New York, pp. 152–67.
2 Pledge, HT (1959) Microscopy, classification, geology. In HT Pledge *Science Since 1500*. Harper, New York, pp. 85–102.
3 Richardson, M (1997) *The Penguin Book of Firsts*. Penguin, Melbourne.
4 Villee, CA (1977) *Biology*, 7th edn. WB Saunders, Philadelphia.
5 O'Brien, TP & McCully, ME (1981) *The Study of Plant Structure — Principles and Selected Methods*. Termarcarphi, Melbourne.
6 Kaplan, DR & Cooke, TJ (1996) The genius of Wilhelm Hofmeister: The origin of causal analytical research in plant development. *American Journal of Botany* 83: 1647–60.
7 Pledge, HT (1959) Cytology and genetics. In HT Pledge *Science Since 1500*. Harper, New York, pp. 218–27.
8 Dunmore, J (1973) Pasteur. In J Dunmore (ed.) *Anthology of French Scientific Prose*. Hutchinson Educational, London, pp. 93–97.
9 Watson, JD (1970) *The Double Helix*. Penguin Books, London.
10 Watson, JD & Crick, FHC (1953) Molecular structure of nucleic acid. A structure for deoxyribose nucleic acid. *Nature* 171: 737–38.
11 Watson, JD & Crick, FHC (1953) Genetic implications of the structure of deoxyribonucleic acid. *Nature* 171: 964–67.
12 Schimmel, P (1987) Aminoacyl tRNA synthetases: general scheme of structure–function relationships in the polypeptides and recognition of transfer RNAs. *Annual Review of Biochemistry* 56: 125–58.
13 Williams, JG & Patient RK (1988) *Genetic Engineering*. IRL Press, Oxford and Washington DC.
14 Howe, CJ & Ward, JS (1989) *Nucleic Acid Sequencing — A Practical Approach*. Oxford University Press, Oxford, New York, Tokyo.

15 Weeden, N F (1991) Chromosomal organization and gene mapping. In DR Murray (ed.) *Advanced Methods in Plant Breeding and Biotechnology*. CAB International, Oxford, pp. 23–49.
16 Stryer, L (1981) *Biochemistry*, 2nd edn. WH Freeman and Company, San Francisco.
17 Breitbart, RE, Andreadis, A & Nadal-Ginard, B (1987) Alternative splicing: a ubiquitous mechanism for the generation of multiple protein isoforms from single genes. *Annual Review of Biochemistry* 56: 467–95.
18 Murray, DR (1999) Glossary. In DR Murray *Growing Peas and Beans*. Kangaroo Press, Sydney, pp. 67–69.
19 de Duve, C (1996) The birth of complex cells. *Scientific American* 274(4): 38–45.
20 Rose, RJ (1991) Modification of the chloroplast genome with particular reference to herbicide resistance. In DR Murray (ed.) *Advanced Methods in Plant Breeding and Biotechnology*. CAB International, Oxford, pp. 222–49.
21 Olmstead, RG & Palmer, JD (1994) Chloroplast DNA systematics: A review of methods and data analysis. *American Journal of Botany* 81: 1205–24.
22 Smith, SE (1989) Biparental inheritance of organelles and its implications in crop improvement. *Plant Breeding Reviews* 6: 361–93.
23 Pelletier, G (1991) Chloroplast and mitochondrial genomes: Manipulation through somatic hybridization. In DR Murray (ed.) *Advanced Methods in Plant Breeding and Biotechnology*. CAB International, Oxford, pp. 201–21.
24 Bradbeer, JW (1988) *Seed Dormancy and Germination*. Blackie, Glasgow and London.
25 Murray, DR (1988) Embryogenesis and prospects for plant improvement. In DR Murray *Nutrition of the Angiosperm Embryo*. Research Studies Press, Taunton, UK, pp. 195–204.
26 Channel 9/WIN Television (2000) 'Who Wants to be a Millionaire?' Melbourne, 7 August.
27 Simmonds, NW (ed.) (1976) *Evolution of Crop Plants*. Longman, London and New York.
28 Gororo, NN, Eastwood, RF, Eagles, HA, Nicolas, ME & Halloran, GM (1999) *Triticum tauschii* wheat derivatives — potential for higher grain yield under moisture deficit. *Proceedings 11th Australian Plant Breeding Conference*, CRC for Molecular Plant Breeding, Adelaide, Volume 2: 180–81.
29 Davidson, BR & Davidson, HF (1993) European settlement in Australia. In BR Davidson and HF Davidson *Legumes: The Australian Experience*. Research Studies Press, Taunton, UK, pp. 166–77.
30 Murray, DR (2000) Fruiting vines and strawberries. In DR Murray *Successful Organic Gardening*. Kangaroo Press, Sydney, pp. 84–86.
31 Murray, DR (1999) Breeding peas and beans. In DR Murray *Growing Peas and Beans*. Kangaroo Press, Sydney, pp. 46–51.
32 Peters, JA (ed.) (1959) *Classic Papers in Genetics*. Prentice-Hall, Englewood Cliffs, New Jersey.
33 Fisher, RA (1936) Has Mendel's work been rediscovered? *Annals of Science* 1: 115–37.
34 Bennett, MD & Leitch, Ilia (2000) Variation in nuclear DNA amount (C-value) in monocots and its significance. In KL Wilson and DA Morrison (eds) *Monocots: Systematics and Evolution*. CSIRO Publishing, Melbourne, pp. 137–46.
35 Bennett, MD (2000) Genomic organisation and systematics in the 21st century. In KL Wilson and DA Morrison (eds) *Monocots: Systematics and Evolution*. CSIRO Publishing, Melbourne, pp. 147–56.
36 Walbot, V (1991) Maize mutants for the 21st century. *The Plant Cell* 3: 851–56.

37 Kyoto Encyclopedia of Genes and Genomes. Bioinformatic Center, Institute for Chemical Research, Kyoto University, Japan, <http://www.genome.ad.jp/KEGG>, accessed 29 May 2002.
38 Coghlan, A (2001) Banana bonanza. *New Scientist* 171(2300): 7.
39 Coghlan, A (2000) Land of opportunity. *New Scientist* 168(2263): 31–33.
40 Spooner, DM, Anderson, GJ & Jansen, RK (1993) Chloroplast DNA evidence for the interrelationships of tomatoes, potatoes and pepinos (Solanaceae). *American Journal of Botany* 80: 676–88.
41 Murray, DR (2000) Fruits as vegetables. In DR Murray *Successful Organic Gardening*. Kangaroo Press, Sydney, pp. 75–80.
42 Robinson, J & Harris, SA (2000) A plastid DNA phylogeny of the genus *Acacia* Miller (Acacieae, Leguminosae). *Botanical Journal of the Linnean Society* 132: 195–213.
43 Maslin, BR (2001) *Wattle, Acacias of Australia*. CSIRO Publishing, Collingwood.
44 Lockhart, PJ & Howe, CJ (2002) *Reconstructing Evolutionary Trees*. BIOS Scientific Publishers, Oxford.
45 Lang, R (2000) Return of the Tassie tiger. *Aussie Post*, 30 December, pp. 14–15.
46 Grimaldi, DA (1996) Captured in amber. *Scientific American* 274(4): 70–77.

2
HOW GENETICALLY MODIFIED PLANTS ARE PRODUCED

> As a cart is to a car, plant breeding is to biotechnology.
> Monsanto[1]

METHODOLOGY AND MYTHOLOGY

This chapter provides a brief overview of the various procedures currently used to introduce new genes into plant cells, leading to the production of transgenic plants. In the popular press there is a great deal of misunderstanding as to what is involved. It is pointless blaming the media. Journalists print what people tell them, and myths soon take hold. Misinformation has often been put forward by proponents of the new technology, as well as by people who have foolishly ventured outside their own areas of expertise. One such is medical Professor John Dwyer, who claims that:

> The voluminous data collected indicate GM crops with major survival advantages that improve agricultural efficiency do not differ from traditionally grown crops except for the presence of the inserted gene.[2]

This single sentence encapsulates the mythology nicely. It contains three glaring inaccuracies. First, the data are sparse rather than voluminous, and not readily available to the public. Regulatory bodies conceal 'commercial in confidence' details, and ignore their implications (Chapter 6). Secondly, transgenic plants can have impaired photosynthetic capacity, and are usually grown at significant extra cost compared with unmodified plants (Chapter 3). Thirdly, genetically engineered plants always differ by more than the single gene they are supposed to receive. The desired gene is always packaged with promoter sequences and reporter genes (see below), and sometimes other genes whose presence is unsuspected. This is what happened with the Showa-Denko attempt to enhance tryptophan production by a genetically modified bacterium in 1989. The gene governing production of a toxic amino acid was unwittingly transferred. The result was a contaminated tryptophan supplement that killed 37 people and left at least 1500 with permanent disabilities. An estimated 5000 people suffered from eosinophilia myalgia syndrome caused by this supplement.[3]

The oft-repeated claim that only a single gene is being transferred has to be set against the knowledge that vectors or constructs used to transfer a desired gene always include several genes. Promoter sequences that respond to signals to switch on the gene and termination sequences that indicate the end of transcription must be included. In addition, reporter genes are included so that cells likely to have integrated the new genes can be identified simply. A construct is generally made to resemble a bacterial plasmid, with all the components arranged in a circle. On cleavage, the resulting linear sequence can be 'spliced' into a cleavage site within chromosomal DNA. Because the different components of a transgene construct usually come from diverse sources, such genes are often termed 'chimeric'. They resemble the Greek mythological figure, the chimera, which had a lion's head, a goat's body, and a serpent's tail.

Reporter genes indicate the presence of the construct in a cell by a colour test, by fluorescence, or by their survival in tissue culture in the presence of an antibiotic or herbicide, when the reporter gene confers resistance to that compound. The most commonly used colour indicator is a synthetic compound that produces a blue pigment following hydrolysis by the newly expressed marker enzyme, ß-glucuronidase, or GUS for short.[4] The gene for this enzyme usually comes from the bacterium *Escherichia coli*.

Another promising reporter system involves the emission of light flashes, as in the abdomen of the firefly (*Photinus pyralis*). The gene for the enzyme luciferase can be transferred either from the firefly or from a bacterium (*Vibrio harveyi*). Assays for light emission are very sensitive, but the equipment required makes this assay relatively expensive.[5] An alternative visual assay that requires only a conventional fluorescence microscope uses the gene for the green fluorescent protein from the jellyfish *Aequorea victoria*.[6]

These reporter systems strongly indicate transformation, but they cannot reveal the location(s) of the introduced genes. Individual transformed plantlets from the same batch can have the transgene package in different locations within the genome. This means that the plantlets are not uniform, unless they all happen to come from a single transformed cell.

Irrespective of which transformation technique is chosen, it is not possible to control or predict where the introduced genes will be located in the recipient genome or how many complete or partial copies will be inserted. This is the antithesis of precision. An example of this kind of imprecision is Monsanto's Roundup Ready soybean, which has now been shown to have one complete copy and two fragments of the bacterial gene specifying the enzyme that allows the plant to modify the herbicide glyphosate (Chapter 3). Scientists working with transformation techniques concede that this imprecision is a problem. In a forum on plants engineered to produce insecticidal proteins from *Bacillus thuringiensis* (Bt proteins):

> Concern was expressed over the lack of control over insertion site in transgenic plants, and the resulting variability. It was acknowledged that transgenesis was largely a 'shot in the dark' and it was pointed out that little was known about the massive variability of individual genotype expression within populations.[7]

There are no guarantees as to the degree of expression of any gene suddenly placed in a new genomic context. Underexpression of the transferred gene is most likely, especially if bacterial genes are placed in a eukaryotic genome. This is a long-standing difficulty.[8] Context for gene expression is extremely important.[9] However, because the insertion of new genes is governed by the specificities of DNA-splicing enzymes and by factors relating to cell type,[10] there are many possible locations for the new gene package. Consequently, the normal expression of pre-existing genes can easily be disrupted.

The insertion of a gene package into one particular chromosome makes that chromosome different from its untransformed homologue. Subsequent cell divisions must be able to accommodate such sequence differences in order to complete successfully. Unexpected effects of gene incorporation are common, often leading to significantly reduced yield. This is a likely consequence of disrupting the functional arrangements of genes that have been selected over countless generations. The interactions of DNA strands with their regulatory histone proteins can also be disturbed by insertion of new gene packages — another possible impact on the utility of the whole genome.

Probe techniques that enable the curious to find out where the introduced genes are located have been developed.[11,12] It is even possible to apply probes to determine the location of the GUS marker gene.[4,13] However, localisation of introduced genes is not always carried out before commercial release of a transformed variety.

The synthesis of any particular protein is subject to delicate and precise controls, governing how much is made, for how long, and in which tissues. Some proteins, such as those stored in seeds, are tissue and organ specific, and produced nowhere else in the plant. Lack of organ specificity in gene expression has been a major problem for genetic engineers. A prime example of this kind of imprecision is the production of a toxic Bt-protein in pollen grains as well as in every other part of transgenic maize plants (Chapter 6).

Sometimes the aim of genetic modification is to inactivate, or 'silence', a gene that is already present. A good example is the blocking of an enzyme necessary for ethylene synthesis. Ethylene is a gaseous plant hormone that promotes senescence, wound healing or fruit ripening. Preventing the synthesis of ethylene is useful for extending the vase-life of cut flowers, as in the 'Moon' series of Florigene carnations,[14] or for delaying the ripening of fruits such as tomatoes. The latter application of gene technology should be much safer than gassing produce with nitric oxide, 'a toxic substance, a contributor to photochemical smog, and a greenhouse gas all in one'.[15]

A plant genome might also be modified by adding a piece of nucleic acid that has been completely synthesised from simple nucleotide precursors. An artificial gene like this can be tailor-made for a specific purpose, for example, it might be an antisense sequence intended to block the local synthesis of ethylene, or it might disrupt the replication of an invading virus. The synthetic gene still has to be packaged in a construct with promoter and reporter genes, so even in cases like these it is never true that only a solitary gene distinguishes a transgenic plant from its unmodified progenitor.

ELECTROPORATION

The possibility that plant cells might take up external DNA and integrate new genes was eagerly anticipated throughout the 1970s.[8,16,17] This prospect moved gradually from the realm of science fiction to science fact thanks to the efforts of many scientists. One key factor was the refinement of techniques for isolating viable protoplasts — plant cells liberated from their surrounding cell walls.

Electroporation is a direct transformation technique that usually depends on plant tissue first being treated to release protoplasts. Under the influence of an electric field, the protoplast's surrounding membrane can form transient pores large enough to admit pieces of DNA — hence the term 'electroporation'. The composition of the surrounding medium and the field strength are two important variables that have to be adjusted experimentally to optimise the results.[5,18]

Once pieces of DNA have entered the protoplast, the procedure relies on chance migration plus enzymic splicing for gene incorporation somewhere in the recipient genome. For successful transformation, the protoplasts must be encouraged to undergo cell division, make new cell walls, and differentiate into plantlets. Some species are particularly well suited to this methodology, for example, lucerne (*Medicago sativa*).[18]

BIOLISTICS

The similarity to the word 'ballistics' is deliberate and appropriate. This is the 'shot-gun' approach, described as 'crude but effective'.[19] Tiny projectiles are coated with DNA preparations before being enclosed in a cartridge case and literally shot into pieces of plant tissue. The procedure poses obvious hazards to the operators, who have to be licensed by the police. One policeman shot himself in the foot while inspecting and testing a primitive machine in South Australia in 1990.[20]

The bombarded plant tissue then has to be coaxed into producing new plants in a suitable culture medium. Transformed plantlets are recovered with the aid of reporter genes for markers such as resistance to the antibiotic kanamycin, mediated by an enzyme activity (phosphate group transfer) under the control of a bacterial gene. When kanamycin is included in the tissue culture medium, cells that have not been transformed will die, leaving only cells that have the highest probability of having integrated the new gene package.

Biolistic procedures have certain advantages. They avoid the need to prepare protoplasts, and provide an opportunity to transform cereals such as wheat, maize and rice, which generally lie outside the normal host range for *Agrobacterium* (see below). Another advantage is that direct transformation of chloroplast or mitochondrial genomes becomes a possibility, because the shot DNA can end up inside an organelle.

As with electroporation, successful incorporation of new genes requires vector delivery, followed by enzyme-catalysed splicing. In contrast to other methods, however, biolistics often results in a large number of gene copies being integrated, with increased risks of disruption to the efficiency of the whole genome. To reduce the number of gene copies incorporated, pre-incubation of tissue with nicotinamide has been shown to be effective (see Table 2.1). Nicotinamide inhibits one of the enzymes involved in repairing breaks in DNA.[21]

In one study,[21] young wheat embryos were separated from developing grain and cultured to encourage callus growth. Bombardment was carried out after incubation of sliced pieces of callus in the presence or absence of nicotinamide. The main aim of the study was to produce wheat plants with nuclear male sterility, so that all the flowers would be functionally female, avoiding the need to remove anthers from unopened flowers by hand. Hybrids might then be produced conveniently just by introducing the desired pollen. The differences between modified and unmodified flowers are illustrated in Figure 2.1.

Selective destruction of tissues responsible for producing pollen grains was accomplished with 'terminator' technology (see also Chapter 4). The bacterial ribonuclease barnase, which degrades all forms of RNA, was specifically expressed in the tapetum, the tissue in the anther from which pollen-mother cells develop. As a precaution against barnase being accidentally expressed in any other part of the plant, expression of the gene that produces the barnase-inactivating compound, bar or barstar, was also engineered. Each of these genes was coupled to a different promoter in the construct: barnase with a tapetal promoter, and bar with a general viral promoter.

Figure 2.1
Wheat florets from a male-fertile control plant and from a transgenic male sterile plant expressing the gene for barnase: A male fertile plant; B male sterile plant; C dismembered male fertile floret just before anther maturity; and D dismembered male sterile floret, showing degeneration of the anthers and swelling of the ovary (ovarium). Pictures courtesy of Dr Marc De Block.

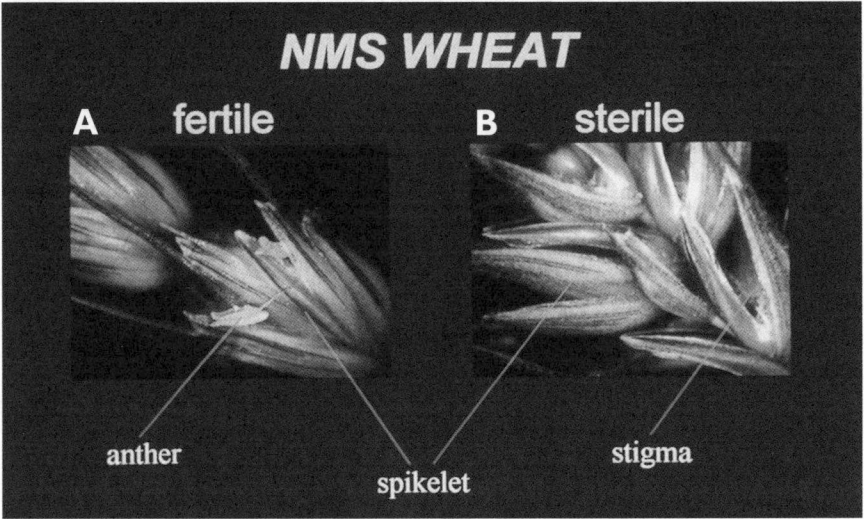
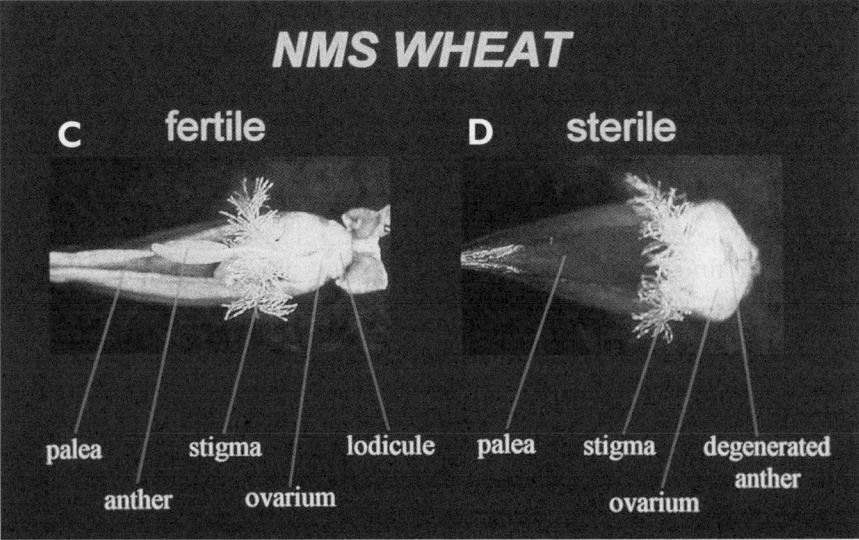

Table 2.1
Influence of a pre-treatment with nicotinamide on transformation frequency and integration of the delivered DNA construct in wheat embryogenic callus tissue[a]

Pre-treatment	Total calli bombarded	Total number transformed	Number of transformants with each copy number for barnase and bar[b]				
			1	2	3	4	5
None	1556	4	**0**	**0**	**0**	**0**	**4**
			0	0	0	0	4
4 days, 2 mM nicotinamide	1653	15	**1**	**4**	**7**	**3**	**0**
			2	4	4	2	3

[a]Data of M. De Block, D. Debrouwer and T. Moens (1997).[22] Wheat embryo calli were bombarded with a construct containing barnase and bar, each with distinct promoters (see text), plus the gene for resistance to the herbicide glufosinate (phosphinothricin) as a selectable marker.
[b]The gene copy number for barnase is given in bold above the number for bar.

The results show that for any individual transformed plant produced following nicotinamide pre-treatment, the number of gene copies integrated for barnase is unlikely to be the same as the number integrated for bar (Table 2.1). A lower gene copy number was obtained at the expense of the association of the main components of the construct, which might in this case assist the non-expression of bar in the tapetum, where barnase is required to act.

It is clear from the background literature referred to by these researchers[21] that terminator technology was used to generate nuclear male sterility well before the notorious proposal to prevent the growth of seedlings from farm-saved seeds (Chapter 4).

INDIRECT METHODS OF GENETIC TRANSFORMATION

In preparing constructs to convey new genes, one option is to use *E. coli* cloning vectors. Another is to use plasmids from a bacterium that is already adapted to infecting plants, namely *Agrobacterium*. This is a genus of soil-dwelling bacteria that cause tumours (crown gall) or hairy roots by transferring plasmids that automatically become integrated with the host plant's genome. The two kinds of plasmid are characterised as Ti (tumour inducing) or Ri (root inducing). In these responses, specific DNA sequences (T-DNA or transferred DNA) are transferred from plasmids into the host genome.[22,23] Following transformation by this bacterium, the plant manufactures hormones that bring about its changed growth pattern, as well as amino acid conjugates (amino acids bound to other molecules, such as a sugar or organic acid)[24] that support the growth of the resident bacteria.

Genes that scientists want to transfer can be placed inside the border regions of Ti or Ri plasmids. Two of the genes normally transferred with these plasmids specify endonucleases capable of cutting the T-DNA and promoting extra synthesis of this excised portion. The *Agrobacterium* system is thus ideal for conveying genes to responsive plants, especially as the plasmids can be 'disarmed' so that the full process of infection does not eventuate.

The modified plasmids can be put back into selected strains of *Agrobacterium*, which are then cultured with some suitable part of the recipient plant. Protoplasts, seeds about to germinate, or discs cut from leaves or other plant parts can all be tested to find out which works best. Transformation frequencies vary from one to 100 per cent.[23] A wide range of plants can be transformed in this way, although wheat and other cereals have usually not been amenable to this procedure until recently.[6]

Viral promoters are often used in gene constructs, but the choice is restricted because most plant viruses possess a core of RNA rather than DNA. Cauliflower mosaic virus (CaMV) has double-stranded DNA, and has provided the most widely used viral promoter sequence (CaMV35S), despite earlier concerns about its limitations.[25] A major problem, however, is the lack of tissue or organ specificity for the expression of genes controlled by this promoter.

The geminiviruses have circular single-stranded DNA surrounded by double (twin) icosahedral clusters of proteins.[26] There are two broad groups: those that are monocotyledon-specific, transmitted by leafhoppers, and those that are dicotyledon-specific, mostly transmitted by whitefly (*Bemisia tabaci*). Two geminiviruses in particular have been developed as vectors for transgenes: tomato golden mosaic virus (TGMV) and cassava latent virus (CLV). One successful method of transformation has involved including sequences from these viruses in Ti plasmids.[26]

TERMINOLOGY

The methods of genetic modification described in this chapter are often collectively termed 'genetic engineering', which for plants can be considered as 'any nonconventional method of genetic manipulation dealing with the transfer of genes between plants and from other organisms to plants'.[17] Another definition is 'the modification of an organism's genetic information by methods other than breeding and selection'.[27] Some people consider these definitions too broad. The authors of this first definition intended to include whole chromosome transfer and cell fusion, which legislation might now seek to exclude. The second definition includes mutagenesis, which in a crude form has been regarded as one of the tools available to the conventional plant breeder. Nevertheless, directed or site-specific mutagenesis very clearly belongs to the field of genetic engineering:

> with the advent of methods for specific cleavage and enzymatic manipulation of DNA, cloning of DNA fragments, nucleotide sequence analysis, and rapid chemical synthesis of oligonucleotides of defined sequence, it is now possible to construct mutations at predetermined sites in a cloned DNA molecule, precisely define the chemical nature of the mutational change, and then test the functional effect in vitro and/or in vivo.[28]

Terms such as 'genetic modification' or 'genetic manipulation' are not at first sight synonymous with 'genetic engineering', since in ordinary conversation they include standard methods of plant breeding. However, attempts to restrict the meaning of 'genetically modified' have now been put into legislation to specify the plants to be regulated and to distinguish them from the products of plant breeding activities that do not warrant such close monitoring.

In the European Community, relevant definitions were set out in Directives in 1990, revised in 1998,[29] as:

> 'organism' is any biological entity capable of replication or of transferring genetic material; 'genetically modified organism (GMO)' means an organism in which genetic material has been altered in a way that does not occur naturally by mating and/or natural recombination.

This definition of 'genetically modified organism' excludes organisms altered by the induction of polyploidy, a technique sometimes used in conventional plant breeding. Agents such as colchicine, an alkaloid from the autumn crocus (*Colchicum autumnale*), can be used for this procedure. During cell division, colchicine stops chromosomal separation by interfering with the contraction of spindle fibres to which the chromosomes are attached.[30] If the colchicine is removed, such polyploid cells often recover and give rise to new plants. When the diploid number of chromosomes is doubled, the resulting plantlets are autotetraploids (4n). Likewise, following tissue culture of anthers, the haploid chromosome number can sometimes be doubled to 2n.[31,32]

Terminology similar to that in the European Directives has been adopted for the Australian Gene Technology Bill (2000). As to exclusions, the Regulations for this Bill (as revised in January 2001)[33] state that:

> the following organisms are not genetically modified organisms:
>
> (a) a mutant organism in which the mutational event did not involve the introduction of any foreign nucleic acid (that is, non-homologous DNA, usually from another species);
>
> (b) a recombinant organism formed by integration, into chromosomal or extrachromosomal DNA sequences, of a genetic element that:

(i) occurs naturally in the species concerned; and

(ii) moves sporadically between genome sites;

(c) an organism that:

(i) results from the fusion of two animal cells; and

(ii) is unable to form a viable whole animal;

(d) an organism that results from protoplast fusion involving only non-pathogenic bacteria or non-pathogenic yeast;

(e) a plant formed by

(i) embryo rescue; or

(ii) in vitro fertilization; or

(iii) zygote implantation; or

(iv) protoplast fusion between sexually compatible species.

It would seem that a plant formed following protoplast fusion between sexually incompatible species will be treated as a genetically modified organism. Because sexual incompatibility provides a reason for attempting this kind of somatic hybridisation, a significant number of genetically modified plants can be expected to arise in this fashion.

Rearrangement of the genome during tissue culture, a phenomenon termed 'somaclonal variation', has been excluded under (a) above, and left in the province of the conventional plant breeder. Once it was widely believed that all cells resulting from mitotic divisions in tissue culture were genetically identical to the original cells, so that all new plantlets differentiating in culture represented one clone. But this assumption was incorrect. The term 'somaclonal variation' was devised by Larkin and Scowcroft[34,35] to account for the heritable variation that they observed in plants derived from tissue culture. Evidently, the process of tissue culture itself causes some rearrangement within the nuclear genome, often allowing the placement of 'silent' genes under different promoters so that they are then expressed.

Somaclonal variation has been particularly useful for revealing heritable resistance to various fungal toxins, permitting resistance to pathogens to be expressed in subsequent progeny. Occasionally a virus-resistant strain results. Disease-resistant varieties of sugar cane, maize, rice, canola, lucerne and tomato have all been produced.[36] This simple procedure is an inexpensive way of generating new plant varieties, and has been widely adopted in India and China.

In practice, the newer techniques of genetic recombination should not be considered in isolation from conventional measures for plant breeding, despite the frequent assumption that the latest techniques stand alone and can deliver marvellous new plant varieties by themselves. Monsanto's claim quoted at the head of this chapter is breathtaking in its arrogance, but typical of the 'revolutionary' view of biotechnology that displays a complete misunderstanding of the context in which genetic engineering has to operate. On the contrary, 'biotechnology alone is unlikely to create useful varieties unless it is combined with a number of procedures used in conventional breeding'.[37] As noted earlier, these supporting procedures include embryogenesis in tissue culture, a technique that has a vital role to play in converting transformed or fused protoplasts into new plants.[38,39]

The integration of old and new techniques, as previously recommended,[40,41] still has the promise to bring about worthwhile improvements in some plant species. However, the goals of genetic modification of plants must be scrutinised much more carefully than has been done so far. The stated goals of people using these new methods are not equally desirable, although farmers have often been exhorted to embrace the new technology without considering all the arguments for and against. The best interests of many farmers lie in protecting niche market opportunities afforded by high quality non-genetically modified crops. Returns for quality produce are often two to three times better than for run-of-the-mill, as organic growers everywhere will attest. Worthwhile breeding goals are considered further in Chapters 4 and 5.

REFERENCES

1 Monsanto (1995) INGARD — some basic facts and perspectives. Monsanto Melbourne.
2 Dwyer, J (1999) Waking up to genetic modification. *Inner Western Suburbs Courier*, 29 November, p. 28.
3 Anderson, L (2000) What is genetic engineering? In Luke Anderson *Genetic Engineering, Food, and our Environment*, Scribe Publications, Melbourne, pp. 19–31.
4 Jefferson, RA, Kavanagh, TA & Bevan, MW (1987) GUS fusions: ß-glucuronidase as a sensitive and versatile gene fusion marker in higher plants. *EMBO Journal* 6: 3901–07.
5 Rathus, C & Birch, RG (1991) Electroporation for direct gene transfer into protoplasts. In DR Murray (ed.) *Advanced Methods in Plant Breeding and Biotechnology*. CAB International, Oxford, pp. 74–102.
6 Weir, B, Gu, X, Wang, M, Upadhyaya, N, Elliott, AR & Brettell, RIS (2001) *Agrobacterium tumefaciens*-mediated transformation of wheat using suspension cells as a model system and green fluorescent protein as a visual marker. *Australian Journal of Plant Physiology* 28: 807–18.
7 Staples, J (1994) Forum report: Evaluation of transgenic plants. In RJ Ackhurst (ed.) *Proceedings of the 2nd Canberra Meeting on* Bacillus thuringiensis. CSIRO Division of Entomology, Canberra, pp. 159–62.
8 Owens, LD (1979) Binding of ColE1-kan plasmid DNA by tobacco protoplasts. *Plant Physiology* 63: 683–86.
9 Suzuki, D & Dressel, H (1999) Unnatural selection. In D Suzuki and H Dressel *Naked Ape to Superspecies*. Allen & Unwin, Sydney, pp. 98–126.
10 Breitbart, RE, Andreadis, A & Nadal-Ginard, B (1987) Alternative splicing: A ubiquitous mechanism for the generation of multiple protein isoforms from single genes. *Annual Review of Biochemistry* 56: 467–95.

11 Wimber, DE & Steffensen, DM (1973) Localization of gene function. *Annual Review of Genetics* 7: 205–23.
12 Mouras, A, Hinnisdaels, S, Taylor, C & Armstrong, KC (1991) Localization of transferred genes in genetically modified plants. In DR Murray (ed.) *Advanced Methods in Plant Breeding and Biotechnology*. CAB International, Oxford, pp. 128–57.
13 Sato, T, Thorsness, MK, Kandasamy, MK, Nishio, T, Hirai, M, Nasrallah, JB & Nasrallah, ME (1991) Activity of an S locus gene promoter in pistils and anthers of transgenic plants. *The Plant Cell* 3: 867–76.
14 Fox, R (1997) Carnations are blue. *Practical Hydroponics and Greenhouses* 33: 38–39.
15 Murray, DR (2000) Plants old and new. In DR Murray *Successful Organic Gardening*. Kangaroo Press, Sydney, pp. 91–93.
16 Chaleff, RS & Carlson, PS (1973) Somatic cell genetics of higher plants. *Annual Review of Genetics* 8: 267–78.
17 Kleinhofs, A & Behki, R (1977) Prospects for plant genome modification by nonconventional methods. *Annual Review of Genetics* 11: 79–101.
18 Larkin, PJ, Taylor, BH, Gersmann, M & Brettel, RIS (1990) Direct gene transfer to protoplasts. *Australian Journal of Plant Physiology* 17: 291–302.
19 Franks, T & Birch, RG (1991) Microprojectile techniques for direct gene Transfer into intact plant cells. In DR Murray (ed.) *Advanced Methods in Plant Breeding and Biotechnology*. CAB International, Oxford, pp. 103–27.
20 Scott, NS (1990) personal communication.
21 De Block, M, Debrouwer, D & Moens, T (1997) The development of a nuclear male sterility system in wheat. Expression of the barnase gene under the control of tapetum specific promoters. *Theoretical and Applied Genetics* 95: 125–31.
22 Zambryski, P (1988) Basic processes underlying *Agrobacterium*-mediated DNA transfer to plant cells. *Annual Review of Genetics* 22: 1–30.
23 Grant, JE, Dommisse, EM, Christey, MC & Conner, AJ (1991) Gene transfer to plants using *Agrobacterium*. In DR Murray (ed.) *Advanced Methods in Plant Breeding and Biotechnology*. CAB International, Oxford, pp. 50–73.
24 Bates, HA (1986) Structure of crown gall tumour metabolites — chemical mediators of parasitism. In R Breslow (ed.) *International Symposium on Bioinorganic Chemistry, Annals of the New York Academy of Sciences* 471: 289–90.
25 Tachini, P & Walbot, V (1986/1987) Transformation of plants. *Nestlé Research News 1986/87*. Vevey, Switzerland, pp. 18–29.
26 Coutts, RHA, Buck, KW & Hayes, RJ (1990) Development of geminivirus-based gene vectors for dicotyledonous plants. *Australian Journal of Plant Physiology* 17: 365–75.
27 Murray, DR (1999) Glossary. In DR Murray *Growing Peas and Beans*. Kangaroo Press, Sydney, pp. 67–69.
28 Shortle, D, DiMaio, D & Nathans, D (1981) Directed mutagenesis. *Annual Review of Genetics* 15: 265–94.
29 European Union (1998) *The CEC Directive on the Release of GMOs* 3 April 1990; *Council Directive* 98/81/EC 26 October 1998.
30 Blakeslee, AF & Avery, AG (1937) Methods of inducing doubling of chromosomes in plants. *Journal of Heredity* 28: 392–411.
31 Blakeslee, AF, Belling, J, Farnham, ME & Bergner, D (1922) A haploid mutant in the Jimson weed, *Datura stramonium*. *Science* 55: 646–47.
32 Bollon, H & Raquin, C (1986/1987) Haplomethods: A tool for crop improvement. *Nestlé Research News 1986/87*. Vevey, Switzerland, pp. 80–91.
33 The Interim Office of the Gene Technology Regulator (2001) *The Commonwealth Draft Gene Technology Regulations 2000*. Canberra.

34 Larkin, PJ & Scowcroft, WR (1981) Somaclonal variation — a novel source of variability from cell cultures for plant improvement. *Theoretical and Applied Genetics* 60: 197–214.
35 Larkin, PJ & Scowcroft, WR (1983) Somaclonal variation and eyespot toxin tolerance in sugar cane. *Plant Cell, Tissue and Organ Culture* 2: 111–21.
36 Murray, DR (1988) Embryogenesis and prospects for plant improvement. In DR Murray *Nutrition of the Angiosperm Embryo*. Research Studies Press, Taunton, UK, pp. 195–204.
37 Sorenson, JC, Provvidenti, R & Munger, HM (1993) Conclusions: Future prospects, strategies and problems. In MM Kyle (ed.) *Resistance to Viral Diseases of Vegetables: Genetics and Breeding*. Timber Press, Portland, Oregon, pp. 255–64.
38 Williams, EG, Verry, IM & Williams, WM (1982) Use of embryo culture in interspecific hybridisation. In IK Vasil, WR Scowcroft and KJ Frey (eds) *Plant Improvement and Somatic Cell Genetics*. Academic Press, New York, pp. 119–28.
39 Parrot, WA, Merkle, SA & Williams, EG (1991) Somatic embryogenesis: Potential for use in propagation and gene transfer systems. In DR Murray (ed.) *Advanced Methods in Plant Breeding and Biotechnology*. CAB International, Oxford, pp. 158–200.
40 Murray, DR (1991) Breeding plants for the 21st century. In DR Murray (ed.) *Advanced Methods in Plant Breeding and Biotechnology*. CAB International, Oxford, pp. 1–22.
41 Kyle, MM (ed.) (1993) *Resistance to Viral Diseases of Vegetables: Genetics and Breeding*. Timber Press, Portland, Oregon.

3
THE HAZARDS OF HERBICIDE–RESISTANT PLANTS

> ... the most frequent application of genetic engineering to plants, that of conferring herbicide-resistance, involves the greatest probability of genetic escape.
>
> David R Murray[1]

BETTER WEED MANAGEMENT?

Most genetically engineered plants produced so far have been modified to be resistant to a specific herbicide. Such modifications are claimed to encourage farmers to use less herbicide, and to simplify weed management. Both of these claims are contradicted by what has actually happened. The use of herbicide-resistant crop plants in the United States has encouraged the increased use of herbicides, and this has led to an increased burden of residues in processed food and fibre products. Glyphosate-resistant soybeans, such as Monsanto's Roundup Ready variety, currently account for about half the soybeans grown in the United States. The application of glyphosate during their cultivation has increased an estimated two- to five-fold.[2,3] The first ever decline in annual sales figures for soy products in Australia was attributed by Sanitarium's spokesperson, Brad Cook, to the debate about these genetically modified soybeans.[4]

Weed management is not better — it is more difficult. It is constrained by the restrictions advised by the herbicide-manufacturing companies and by the increased recording and reporting requirements of new legislation governing pesticide use. Herbicide tolerance is highly qualified because of the frailties of the plants at certain stages of growth and the likelihood of yield reduction. The yield penalty that accompanies the cultivation of herbicide-resistant crop plants ranges between 10 and 30 per cent,[2,5] and is rarely factored into a farmer's calculations in advance. Siren, the predominant triazine-resistant variety of canola (*Brassica napus*) grown in Australia throughout the 1990s, has a yield penalty of between 15 and 20 per cent.[6]

Large-scale cropping encourages aerial application of herbicides (and pesticides). However, much of what is sprayed is wasted. Spray drift accompanies aerial application of herbicides even under the best climatic conditions, hence the introduction of a drift-retardant formulation, Roundup Max, in 2001. According to Jo Immig of the Total Environment Centre, Sydney,[7] 'Spray drift should essentially be regarded as "chemical trespass" and treated like any other trespass onto your property or person.' Furthermore, the Australian Government's Select Committee Inquiry into Agricultural and Veterinary Chemicals[8] recommended that because of inherent difficulties in implementing uniform legislation, '... aerial spraying should be banned or phased out in Australian Agriculture.'

Some companies, such as Pioneer Hi-Bred, avoided the controversy surrounding genetic engineering techniques by using conventional breeding procedures to develop herbicide-resistant crop plants. Triazine-resistant canola was developed by conventional breeding methods,[9] and the resistance gene has been transferred widely within *Brassica* using embryo rescue or somatic embryogenesis.[10,11] But the deleterious aspects of conferring herbicide resistance are the same. Herbicide-resistant crops provide the rationale for an increase in herbicide application, because any pre-emergent application (made around the time of sowing) is no longer the only application that can be made before harvest. A conservative estimate of the increased herbicide application that results is three-fold.[12]

WHAT ARE HERBICIDES?

The term 'herbicide' had been coined by the mid-1920s to mean compounds that kill plants, usually with some degree of selectivity. Methyl bromide made its debut in 1894. This general toxin could be used to fumigate the soil, killing all life forms. It has lingered as a nematicide (a compound that kills nematode worms), but is now being phased out worldwide. The last bastion of use is in South Australia, where it is still favoured for use on lawns being converted into 'organic' gardens[13] and for keeping sultanas free of insects.

In the late 19th and early 20th century, spraying with sulfuric acid or cyanide was also tried. These universal poisons have obvious drawbacks. The use of chlorates (potassium or sodium salts) began in Queensland in 1901. Potassium chlorate was used against prickly pear

(*Opuntia stricta*) with limited effect. In the 1920s, chlorates were much more successful against Canada thistle (*Cirsium arvense*), and even at rates of 200 to 300 kilograms per hectare (in winter) did not harm earthworms, soil protozoa or subsequent oat crops.[14] The same cannot be said of more recent herbicides, despite up to 1000-fold reductions in the absolute amounts applied per unit area. Residence times of two or three years in soil are common, and accumulation of herbicides in ground water is a major environmental and health concern.[12,15,16]

In the 1940s, the first of the phenoxyacetic acid group was trialled successfully: 2,4-dichlorophenoxyacetic acid (2,4-D). This compound proved to be selective against broad-leafed weeds (i.e. dicotyledons), whereas many grasses and cereals (monocotyledons) were tolerant. 2,4-D and related compounds upset normal growth and development patterns by mimicking the plant hormone auxin (indoleacetic acid), leading to membrane disturbances and the collapse of overly extended cell walls. At suitably low concentrations, 2,4-D can be used as an auxin analogue in media formulated for plant tissue culture. A combination of 2,4-D and 2,4,5-T (2,4,5-trichlorophenoxyacetic acid) known as Agent Orange was produced by Monsanto and used as a defoliant over an area of 1.7 million hectares in Vietnam.[17]

More than 245 herbicides were in common use by 1976.[18,19] Variously substituted methylureas and dimethylureas form one large group. These compounds act primarily by interfering with the light-capturing reactions of photosynthesis. Diuron, or DCMU (dichlorophenyl dimethyl urea) became an important research tool in the 1960s, because its site of inhibition in photosynthesis is so specific.

The triazines, including simazine and atrazine, form another important group. They also interfere with photosynthesis, binding specifically to a small protein (of molecular weight 32,000) encoded by the chloroplast DNA.[5,20] However, at very low concentrations, triazines have growth stimulating effects like another class of plant hormones, the cytokinins.[21]

Glyphosate, or more precisely *N*-(phosphonomethyl) glycine, has been in use for more than 25 years. By inhibiting a specific enzyme, glyphosate blocks the synthesis of aromatic rings, present in the amino acids phenylalanine and tyrosine, plus a host of polyphenolic substances. This primary effect of glyphosate is not confined to plants. However, glyphosate does have the advantage of breaking down quickly in the soil to harmless products, including inorganic phosphate and glycine, the simplest amino acid. This breakdown is conducted by certain soil bacteria such as *Salmonella typhimurium*, which has now provided the gene for the enzyme that allows genetically engineered glyphosate-resistant plants to detoxify some of the glyphosate they are exposed to.[22]

The sulfonylureas (for example, chlorsulfuron) were first marketed in 1982.[15] They prevent the synthesis of amino acids with branched carbon chains, as detailed below. A relative newcomer, glufosinate (phosphinothricin) has brand names Basta and Liberty. Glufosinate inhibits the enzyme glutamine synthetase, thereby preventing the conversion of ammonium ions to amino groups of amino acids. An accumulation of ammonium ions can be toxic to plant cells. Naturally resistant species, such as barley (*Hordeum vulgare*), adapt their biosynthetic pathways by reducing the synthesis of glycine and serine, but increasing the synthesis of many other kinds of amino acid, including those with a branched chain or an aromatic ring.[23]

Herbicides are now classified into mode-of-action groups by letter of the alphabet (Table 3.1). This classification has been necessitated by the appearance of herbicide-resistant weeds. Farmers using herbicides are encouraged not to use the same group constantly, but to use a rotation of herbicides from different groups, as well as physical methods of control.[24,25,26] Variation across groups is recommended because weeds developing resistance to a given herbicide often display co-resistance to other members of the same group, whereas such plants will probably not immediately possess resistance to a herbicide with a sharply different mode of action. The advent of herbicide-resistant weeds is already a major impediment to the continued use of herbicides (see Herbicide-Resistant Weeds below).

Table 3.1
Examples of herbicides listed by alphabetical group

Group	Names
A	diclofop-methyl
B	chlorsulfuron, imidazolinones
C	triazines, methylureas, acetazolamide
D	trifluralin
E	thiocarbamate
F	amitrole, norflurazon
G	azafenidin (triazolone)
H	thiobencarb
I	2,4-dinitrophenoxyacetic acid (2,4-D)
J	flupropanate
K	metolachlor
L	diquat, paraquat
M	N-(phosphonomethyl) glycine (glyphosate)
N	glufosinate (phosphinothricin)

ADVERSE EFFECTS OF HERBICIDES ON NON-TARGET ORGANISMS

Apart from possible effects on people applying sprays or caught in drift, the active ingredients of many herbicides can poison soil microorganisms such as mycorrhizal fungi, and earthworms, beneficial insects, birds and animals.[3,17] But the most dangerous components of herbicide formulations are not necessarily the herbicidal ingredients. The devastating intergenerational deformities produced by exposure to Agent Orange were mediated by tetrachlorodibenzodioxins (dioxins), neglected by-products of organochlorine herbicide manufacture. The effects of such compounds were not acknowledged for a long time,[27] and at one stage it was claimed that they formed only under adverse storage conditions (hot galvanised iron sheds). Alternatively, their presence was admitted, but stories about their breakdown 'in a day' after spraying were promulgated instead.[28] In fact, dioxins can be detected 10 years or more after spraying.[17] Premature deaths among people who sprayed 2,4,5-T without adequate protection in the 1970s and 1980s are the subject of ongoing inquiries.

Ingredients added to increase the effect of the herbicidal ingredient can have serious effects alone, or by acting synergistically with the active ingredients. For example, surfactants are often added to help the herbicide penetrate the wax coating over the epidermal cells of the leaf.[29,30] Earlier formulations of glyphosate had devastating effects on frogs until the surfactant was changed.[31]

Some herbicides give the appearance of being relatively safe. Sulfonylureas, for instance, inhibit an enzyme that is non-functional in humans. This is acetolactate synthase, which catalyses the first step in the synthesis of the branched-chain amino acids found in polypeptides. Because we lack a functional form of this enzyme, all three branched-chain amino acids, valine, leucine and isoleucine, are essential in the human diet. But it is not known whether sulfonylureas have other effects that have not yet been identified.

The synthesis of aromatic rings is another metabolic pathway that humans do not possess, so glyphosate cannot affect an enzyme that we lack and which is crucial to this pathway. However, we do not know whether glyphosate is safe or not. No-one can give an unequivocal assurance about the effects of any particular herbicide on human health. Even in the absence of immediate (acute) effects, it might take 40 years for a potential carcinogen to act in enough people for it to be detected as a cause. Removal of chemicals from the list of those formerly approved is occurring all the time with the benefit of hindsight. For example, triazines are now considered to be endocrine disruptors, and have been banned in France.[16] Accordingly, precautions should always be taken to minimise exposure to any agricultural chemical.

INSECTICIDES AS HERBICIDES

Just as the deleterious effects of herbicides are not confined to plants, so other pesticides can incidentally diminish plant growth. The persistent organochlorine pesticide DDT, which was widely used against insects after World War II, had many adverse effects on non-target species.[32,33] In the 1960s, DDT was shown to have herbicidal effects on cereals, especially barley, rye and wheat. Some varieties were more susceptible than others, and this was linked to possession of a dominant allele at a specific gene locus. Two sites of inhibition of photosynthesis were eventually characterised.[34] These sites of inhibition are distinct from those where Group C herbicides act. In addition, DDT inhibits translocation of sugar from the leaves.[35] How many other pesticides still in common use are also herbicides, and what hidden yield reductions have conventional farmers been putting up with?

SYSTEMIC HERBICIDES

Following uptake through the leaves or roots, herbicides may move from cell to cell, and then be carried in the vascular tissues of the plant. The xylem transport system consists of open capillaries, with thickened cell walls. Transport in the xylem is upwards, towards the leaves. Xylem sap rises in response to the loss of water vapour through the open pores of the leaves (transpiration). Phloem transport cells are much smaller than xylem, and still living. Their end-walls are punctuated with obvious openings, giving rise to the terms 'sieve-plate' for the end-wall, and 'sieve tube member' for the cell that carries out translocation. The phloem sieve-tubes normally carry sucrose, amino compounds, organic acids and mineral ions in both directions, and hence to all non-transpiring parts of the plant, such as roots, shoot tips, buds and seeds. If herbicides are transferred into the phloem, they will be carried to every part of the plant. Herbicides (or other pesticides) distributed throughout a plant via the phloem are described as 'systemic'. Some examples are listed in Table 3.2.

Table 3.2
Examples of herbicide mobility in phloem

Herbicide	Experimental plant	Mobility	Reference
Glyphosate	sugar beet	strong	Gougler[36]
Glyphosate	broad bean	strong	Groussol[37]
2,4-D	broad bean	strong	Groussol[37]
Chlorsulfuron	garden pea	moderate	Murray[38,39,40]
Glufosinate	soybean	moderate	Shelp[41]
Azafenidin	several weeds	weak	McQuinn[42]

Phloem mobility of a herbicide is an advantage so far as killing weeds is concerned. Weeds can be tackled even when they have begun to flower, because herbicide entering through the leaves will be carried into the developing seeds. A non-systemic herbicide applied too late could precipitate seed maturation and shedding, so adding to the soil seed bank. Consider the vine *Anredera cordifolia*, which has aerial tubers. If the stem is severed, the tubers will be shed, and the dying plant will reproduce. But if the stem is nicked and glyphosate applied, it will be transported into the tubers. The chances of killing most of the tubers are excellent. This example illustrates why glyphosate has become indispensable for bush regenerators.

However, phloem mobility is a liability so far as herbicide-resistant crop plants are concerned. Despite many years of uncertainty over the relative contributions of xylem sap and phloem sap to the nutrition of developing fruits, it has now been established that the phloem is the predominant supplier of all nutrients to fruits and their enclosed seeds.[38,40,43,44] Inevitably, a systemic herbicide will be carried into the harvested parts of plants. This was demonstrated in the study with chlorsulfuron included in Table 3.2. At early stages of seed development in peas, the imported nitrogen-rich solute asparagine is converted mainly into glutamine, alanine and valine, which are all secreted into the embryo sac and taken up by the embryo. When a one micromolar solution of chlorsulfuron was supplied via the cut stem, valine synthesis in the seedcoats was substantially inhibited, and increased amounts of other amino acids were produced instead. These observations confirmed that chlorsulfuron can be carried into the seedcoats, acting there to diminish the synthesis of branched-chain amino acids.

It follows that the transmission of herbicides in phloem must result in significant accumulation of herbicides inside the seeds. Normally one would expect seeds protected by pod walls or bracts to acquire very little adventitious herbicide. Thus the phloem mobility of systemic herbicides is the underlying reason why the tolerance for glyphosate in soybeans imported by Australia was raised an enormous 200-fold: from 0.1 milligrams per kilogram to 20 milligrams per kilogram. In this way, Roundup Ready soybeans from the United States were not refused entry, despite massive contamination by former standards. Exactly how much of any particular herbicide is present in the seeds of herbicide-resistant crop plants has not been revealed. Grain products are not included in conventional market surveys for pesticide residues and cadmium.[45]

Systemic herbicides affecting amino acid metabolism will also have an impact on seed protein composition. Grain from herbicide-resistant crop plants is unlikely to match the quality of non-genetically modified standard cultivars. Deficiencies in essential amino acids such as tyrosine, phenylalanine and tryptophan can be expected for glyphosate-resistant soybeans. Likewise, deficiencies in the sulfur-containing amino acids related to glycine and serine (cysteine and methionine) can be anticipated for barley sprayed with glufosinate. Grain depleted of any essential amino acid is not worth as much as grain with normal protein content and amino acid composition. So much for the improved agricultural 'efficiency' of herbicide-resistant crop plants — lower quantity and lower quality.

HERBICIDE-RESISTANT WEEDS

The transfer of herbicide-resistance genes from herbicide-resistant crop plants to weedy relatives was once a major concern. There is abundant evidence that this concern was well justified, but there is now an even greater concern: the selection of herbicide-resistant weeds simply from repeated application of the same herbicide. 'Intensive use of a given herbicide can act as a selective agent in favour of herbicide-resistant genotypes, until the herbicide is no longer useful.'[1] This can act to the advantage of the original manufacturer. A guaranteed obsolescence, timed to coincide with the expiry of patents, would give no advantage to copycat manufacturers.

September 2000 was the significant date for the expiry of patent protection of glyphosate. This was supposed to be a herbicide that could be sprayed without eliciting resistant genotypes in the target weed populations. But in the late 1990s, several examples of glyphosate-resistant annual rye-grass (*Lolium rigidum*) were identified. These appeared after 15 to 30 applications of glyphosate in the same locations, with no other herbicides or methods of weed management involved.[46,47] Altogether, 23 weed species have acquired herbicide resistance in Australia, and this has occurred in advance of the release of any genetically modified herbicide-resistant crop plants. Most instances involve herbicides from Groups A and B,[25] and some have become prominent following very few annual applications of herbicide. For example, annual rye-grass developed sulfonylurea resistance after only four applications.[48] World-wide registers of the appearances of herbicide-resistant weed species are now being kept (Figure 3.1).[49]

MULTIPLY-RESISTANT CANOLA

Canola (*Brassica napus*) is a cultivated plant that is also a weed. The seed is easily lost just before or during harvest, so canola is automatically a weed of subsequent crops grown in the same places, such as beans or wheat. Shed seeds can germinate over a period of two years. Canola is also a weed of the railway and the roadside.

Figure 3.1
Incidence of herbicide-resistant weeds as monitored by Dr Ian Heap[49]

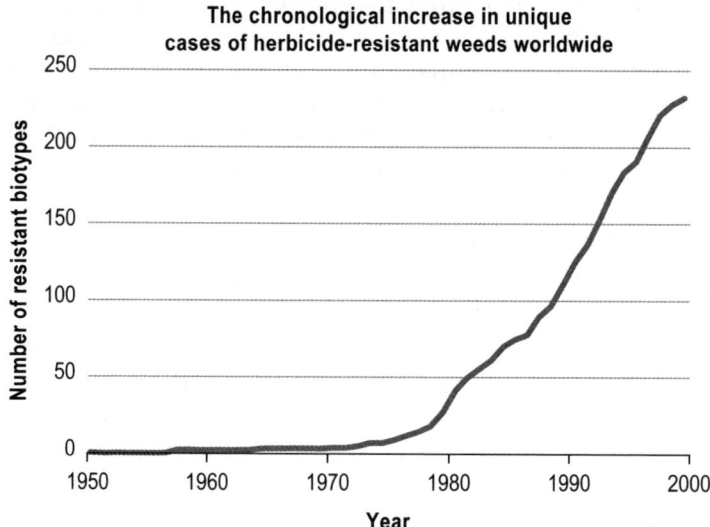

The situation with canola is becoming increasingly complex. Despite assurances from various companies that herbicide-resistant canola cannot cross with non-resistant varieties, there is evidence that this does happen. Different species of *Brassica* can cross under certain conditions[9,10,11] and canola even crosses readily with a weedy descendant of one of its original parents (*B. campestris*, Table 1.1).[50]

A non-resistant variety of canola can readily acquire a single resistance encoded by a gene in the nucleus. This happened to Percy Schmeiser's canola plants in Saskatchewan, Canada. This farmer grew canola and saved his own seed over many years, then found that about 60 per cent of his crop had acquired resistance to glyphosate.[51] The rapid development of triply resistant canola has now been reported in Canada. Neighbouring farmers who grew glyphosate-resistant, glufosinate-resistant or imidazolinone-resistant varieties of canola found that a three-way resistant genotype developed by itself in just two years.[52] Such is the efficacy of insect-mediated crossing. A plant breeder attempting to deliberately 'stack' genes would be delighted with this rate of progress.

It is abundantly clear that once a herbicide-resistant variety is released, its originators have no control over the spread of the resistance gene. Surrounding fields with buffer zones of unmodified crop plant is no impediment, and simply an encouragement to crossing. Arguments about the dimensions of these buffer zones are pointless, as it has long been known that distances of 3.2 to 6.4 kilometres are necessary to prevent unwanted insect-mediated crossing.[53]

NEW HERBICIDES

As weeds develop resistance to herbicides currently in use, there is clearly a financial incentive for companies to spend money on the research and development necessary for the release of new herbicides (Table 3.3).

Table 3.3
New herbicides due for release in 2001

Company	Trade name	Active ingredient	Group	Usual rate (g/ha)
Aventis	Hussar	iodosulfuron	B	50
Dupont	Milestone[a]	azafenidin[b]	G	500–800
Cyanamid	On Duty	imazapic + imazapyr	B	28
Cyanamid	Midas	On Duty + MCPA	B	36

[a] Also Evolus and Galaxy.
[b] A complex organochlorine triazolone molecule with components related to 2,4-D.

Iodosulfuron is a new sulfonylurea herbicide, effective against a wide range of both grass and broad-leaf weeds in wheat crops, but considered too damaging for use on barley.[54] As iodosulfuron and two of the other new releases all belong to Group B, it might be anticipated that co-resistance will be displayed by weeds already resistant to earlier members of Group B. This may well shorten the time over which these new herbicides might be reasonably effective.

In contrast, azafenidin inhibits porphyrin ring synthesis, reducing the ability of a plant to produce essential pigments such as chlorophyll and cytochromes. It kills many annual weed species, and is being recommended for use with grapevines, citrus, pine and eucalypt trees, and sugar cane. Acting as a residual herbicide, it retains activity in the soil for about three months, and will not leach readily. Its breakdown depends on soil bacteria. Consistent with its weak phloem mobility, residues have not been detected in harvested grapes or citrus fruits.[42]

Plant products are also being considered for use as herbicides. For example, the tree of heaven (*Ailanthus altissima*) secretes a compound from its roots called ailanthone, which inhibits seedling growth or seed germination of a wide range of plant species.[55]

ALTERNATIVES TO USING HERBICIDES

Weed management is often equated with deciding which herbicide to spray. This simplistic approach is part of the 1950s mind-set that believes there is a chemical solution to every biological problem. But there are many alternatives to using herbicides, and these are espoused by various codes of organic or ecologically sustainable farming practice.

A great step forward would be to stop planting weeds. Crop seeds purchased for planting are often contaminated with seeds of weeds or other plants. It may take very few seeds for a weed to gain a new focus for distribution, for example parthenium weed (*Parthenium hysterophorus*) in Queensland, where under ideal growing conditions a single plant can quickly generate 15 000 new seeds. Fodder purchased for stock animals can also be a major source of weeds. In Yass (NSW), grain imported over a two-year period contained an average 555 weed seeds per kilogram, and hay fodder imported in the same interval contained 68 700 weed seeds per bale.[26] Even worse is to plant a crop contaminated with seeds of herbicide-resistant weeds. Group A herbicide-resistant wild oats (*Avena* species) have been spread in crop seeds in northern NSW.

Rotation of a field from crop to pasture legume, such as lucerne, forces many weeds to stand out above the groundcover. They can be lopped before flowering. Unfortunately, this is often left too late, and the lucerne is harvested for sale replete with weed seeds. This has led to lucerne mulch having a poor reputation by comparison with weed-free alternatives.[56] Silage is a better option, as any seeds present do not survive.

Zero tolerance of contamination is not too arduous a precaution. This condition already applies to barley, which must be absolutely free of ergot-infested grains. Farmers need to take action against suppliers of weed-infested seed.

Physical methods of weed control are selective, and leave no chemical residues. Steam is making a long overdue comeback, with several propane gas-fuelled appliances now entering the market.[57] Demonstrations are essential, because some machines do not work as well as others. Brief heat treatments can kill plants because they disrupt the phloem cells. Note, however, that perennial grasses are more likely than softer annuals to recover from a single heat treatment.[26]

Cultivation, if it can be timed to avoid erosion by wind or water, is the simplest physical method for aerating the soil and simultaneously destroying weeds. It can be applied wholesale, or selectively if it is thought inadvisable to turn over the whole paddock at once. Although cultivation is out of favour with 'no till' advocates, careful choice of direction, vehicle and plough make a crucial difference to its usefulness. The improved design of the Yeomans plough[58] compared with the 'chisel' nosed alternative ensures that the soil is not recompacted by the vehicle propelling the plough. Double cultivation turns in the weeds that result from seeds brought near the surface with the first cultivation, further reducing the soil seed bank.

Inter-plant competition is under-utilised as a weed control measure. Planting densities should be chosen to give the crop canopy the best possible chance of enclosing and out-competing any weeds. Both wild oats (*Avena fatua*) and *Phalaris paradoxa* can be eliminated from wheat crops by increasing the planting density.[26,59] If weed cover is of the order of 10 per cent or less, there is unlikely to be a pronounced impact on yield. In fact, some groundcover from adjacent weeds is beneficial for crop plant root systems, as the soil is protected against overheating and against erosion by wind or rain. Groundcover from legumes like clovers (*Trifolium* species), as recommended long ago by William Farrer, adds to available nitrogen, contributing positively to yield. Considering the financial return on a crop, money not spent on herbicides and their application (usually by subcontractors) is money added to the profit margin.

CONCLUSION

The adoption of herbicide-resistant crop plants is a move in the wrong direction. Growing such plants encourages greater use of herbicides, and hinders the effective management of weeds by complicating options for the farmer. The suppliers of seeds of herbicide-resistant plants are imposing restrictive conditions in their 'technology use agreements' that include nominating the brand of herbicide to be applied, and prohibiting the saving of seeds for replanting. This creates major problems for growers. Such restrictive trade practices ought to draw the attention of the Australian Competition and Consumer Commission (ACCC).

Spraying herbicides after seedling emergence, when the crops of neighbouring farms could suffer damage from spray drift, is antisocial to say the least, and poses problems of yield loss through inhibition of flowering, contamination of produce and litigation over claims for damages. Increased reliance on herbicide-resistant crop plants is contrary both to modern concepts of integrated pest management, and to the Australian National Weeds Strategy, which seeks to preserve biodiversity in accordance with United Nations conventions.[60] Accumulation of larger quantities of systemic herbicides in grain products is also at odds with consumer expectations of *reduced* herbicide and pesticide residues in foods, rather than the substantial increases that are now occurring.

REFERENCES

1 Murray, DR (1991) Breeding plants for the 21st century. In DR Murray (ed.) *Advanced Methods in Plant Breeding and Biotechnology.* CAB International, Oxford, pp. 1–22.
2 Benbrook, C (1998) Evidence of the magnitude and consequences of the Roundup ready soy bean yield drag from university-based varietal trials in 1998. Benbrook Consulting Services, Sandpoint, Idaho.
3 Anderson, L (2000) Genetic engineering and the environment. In Luke Anderson *Genetic Engineering, Food and our Environment.* Scribe Publications, Melbourne, pp. 33–53.

4   ABC Radio (2000) 'The Country Hour'. 26 October.
5   Rose, RJ (1991) Modification of the chloroplast genome with particular reference to herbicide resistance. In DR Murray (ed.) *Advanced Methods in Plant Breeding and Biotechnology*. CAB International, Oxford, pp. 222–49.
6   Salisbury, PA, Wratten, N, Potter, TD & Burton, WA (1999) Canola — an Australian success story. *Proceedings 11th Australian Plant Breeding Conference*, CRC for Molecular Plant Breeding, Adelaide, Volume 1: 77–80.
7   Immig, J (1999) *Clearing the Air — Pesticide Spray Drift*. Total Environment Centre, Sydney.
8   Australian Federal Government's Select Committee Inquiry into Agricultural and Veterinary Chemicals (1990) Canberra.
9   Beversdorf, WD, Weiss-Lerman, J, Erickson, LR & Souza Machado, V (1980) Transfer of cytoplasmically inherited triazine resistance from bird's rape to cultivated oil seed rape (*Brassica campestris* and *Brassica napus*). *Canadian Journal of Genetics and Cytology* 22: 167–72.
10 Ayotte, R, Harney, PM & Souza Machado, V (1989) The transfer of triazine resistance from *Brassica napus* L to *B. oleracea* L: IV. Second and third backcrosses to *B. oleracea* and recovery of an 18-chromosome, triazine-resistant BC3. *Euphytica* 40: 15–19.
11 Pelletier, G (1991) Chloroplast and mitochondrial genomes: Manipulation through somatic hybridization. In DR Murray (ed.) *Advanced Methods in Plant Breeding and Biotechnology*. CAB International, Oxford, pp. 201–21.
12 Fagan, J (1995) *Genetic Engineering: The Hazards — Vedic Engineering: The Solutions*. Maharishi International University Press, Fairfield, Iowa.
13 Bennett, P (1999) *Organic Gardening*, 6th edn. New Holland, Sydney.
14 Aslander, JA (1928) Experiments on the eradication of Canada thistle, *Cirsium arvense*, with chlorates and other herbicides. *Journal of Agricultural Research* 36: 915–34.
15 Ferris, IG (1993) A risk assessment of sulfonylurea herbicides leaching to groundwater. *AGSO Journal of Australian Geology and Geophysics* 14: 297–302.
16 Sutherland, S (2001) France bans herbicide responsible for water pollution. *A Good Weed* 24: 8.
17 Westing, AH (1977) Ecological effects of the military uses of herbicides. In FH Perring and K Mellanby (eds) *Ecological Effects of Pesticides*. Academic Press, London, pp. 89–94.
18 Dodge, AD (1977) The mode of action of well-known herbicides. In NR McFarlane (ed.) *Herbicides and Fungicides*. The Chemical Society, London, pp. 7–21.
19 Fryer, JD (1977) Recent developments in the agricultural use of herbicides in relation to ecological effects. In FH Perring and K Mellanby (eds) *Ecological Effects of Pesticides*. Academic Press, London, pp. 27–45.
20 Arntzen, CJ, Pfister, K & Steinback, K (1982) The mechanism of chloroplast triazine resistance: Alterations in the site of herbicide action. In H LeBarron and J Gressel (eds) *Herbicide Resistance in Plants*. John Wiley & Sons, New York, pp. 185–214.
21 Nadar, HM, Clegg, MD & Maranville, JW (1975) Promotion of sorghum callus growth by the s-triazine herbicides. *Plant Physiology* 56: 747–51.
22 Comai, L, Facciotti, D, Hiatt, WR, Thompson, G, Rose, RE & Stalker, DM (1985) Expression in plants of a mutant *aro*A gene from *Salmonella typhimurium* confers tolerance to glyphosate. *Nature* 317: 741–44.
23 Shelp, BJ, Swanton, CJ, Mersey, BG & Hall, JC (1992) Glufosinate (phosphinothricin) inhibition of nitrogen metabolism in barley and green foxtail plants. *Journal of Plant Physiology* 139: 605–10.
24 Walker, S (1998) Weed update. *Agdex*, No. 682. DPI Queensland, Dalby.

25 Storrie, A (2000) *Weed Control in Summer Crops 2000–2001*. NSW Agriculture, Dubbo.
26 Storrie, A (2000) Enhanced competition/hygiene in cropping/pasture areas. In *Weed Management 2000*. The Weed Society of NSW, Sydney, pp. 5.1–5.8.
27 Beder, S (1997) Scientific controversy: Dioxin. In S Beder *Global Spin — The Corporate Assault on Environmentalism*. Scribe Publications, Melbourne, pp. 141–60.
28 Boden, A (1979) Industrial and social risks associated with pesticides. In FWG White (ed.) *Scientific Advances and Community Risk*. Australian Academy of Science, Canberra, pp. 125–39.
29 Kirkwood, RC (1977) Some criteria determining penetration and translocation of foliage-applied herbicides. In NR McFarlane (ed.) *Herbicides and Fungicides*. The Chemical Society, London, pp. 67–80.
30 Hamilton, RJ (1977) Plant waxes and penetration of 2,4-dichlorophenoxyacetic acid. In NR McFarlane (ed.) *Herbicides and Fungicides*. The Chemical Society, London, pp. 81–92.
31 Mann, RM & Bidwell, JR (1999) The toxicity of glyphosate and several glyphosate formulations to four species of southwestern Australian frogs. *Archives of Environmental Contamination and Toxicology* 36: 193–99.
32 Carson, R (1962) *Silent Spring*. Houghton Mifflin, Boston, Massachusetts.
33 Graham, F Jr (1972) *Since Silent Spring*. Pan/Ballantine, London.
34 Akbar, S & Rogers, LJ (1985) Effects of DDT on photosynthetic electron flow in *Secale* species. *Phytochemistry* 24: 2785–89.
35 Akbar, S & Rogers, LJ (1985) Effects of DDT on carbohydrate metabolism and translocation in *Secale* species. *Phytochemistry* 24: 2791–95.
36 Gougler, JA & Geiger, DR (1981) Uptake and distribution of N-phosphonomethyl glycine in sugar beet plants. *Plant Physiology* 68: 668–72.
37 Groussol, J, Delrot, S, Caruhel, P & Bonnemain, J-L (1986) Design of an improved exudation method for phloem sap collection and its use for the study of phloem mobility of pesticides. *Physiologie Vegetale* 24: 123–33.
38 Murray, DR (1987) Nutritive role of the seedcoats in developing legume seeds. *American Journal of Botany* 74: 1122–37.
39 Murray, DR (1988) The nutritive function of seedcoats. In David R Murray *Nutrition of the Angiosperm Embryo*. Research Studies Press, Taunton, pp. 121–52.
40 Murray, DR (1990) Acquisition, transformation and secretion of nitrogenous solutes by seedcoats of developing pea seeds. In MA Beilby, NA Walker and JR Smith (eds) *Membrane Transport in Plants and Fungi*. University of Sydney, Sydney, pp. 479–81.
41 Shelp, BJ, Swanton, CJ & Hall, JC (1992) Glufosinate (phosphinothricin) mobility in young soybean shoots. *Journal of Plant Physiology* 139: 626–28.
42 McQuinn, D (2000) Milestone/Evolus — azafenidin (R6447): A new residual herbicide for grapevines, citrus, eucalypts, *Pinus* and sugar cane. In *Weed Management 2000*. The Weed Society of NSW, Sydney, pp. 2.1–2.4.
43 Murray, DR (1988) Translocation mechanisms. In David R Murray *Nutrition of the Angiosperm Embryo*. Research Studies Press, Taunton, UK, pp. 71–93.
44 Murray, DR (1992) Amino acid and amide metabolism in the hulls and seeds of developing fruits of garden pea, *Pisum sativum* L. V. Aspartate. *New Phytologist* 120: 259–64.
45 Plowman, T, Ahmad, N & Bower, C (1998) *Monitoring Pesticide and Cadmium Residues in Fresh Fruit and Vegetables* 1992–5. Horticultural Research and Development Corporation and NSW Agriculture, Orange.
46 Powles, SB, Lorraine-Colwill, DF, Dellow, JJ & Preston, C (1998) Evolved resistance to glyphosate in rigid ryegrass (*Lolium rigidum*) in Australia. *Weed Science* 46: 604–07.

47 Preston, C (1999) Glyphosate resistance in weed species. *A Good Weed* 17: 5–6.
48 Gill, GS (1995) Development of herbicide resistance in annual ryegrass populations (*Lolium rigidum* Gaud.) in the cropping belt of Western Australia. *Australian Journal of Experimental Agriculture* 35: 67–72.
49 Heap, I. International Survey of Herbicide Resistant Weeds. <http://www.weedscience.com>, accessed 2000.
50 Jorgensen, RB & Andersen, B (1994) Spontaneous hybridization between oilseed rape (*Brassica napus*) and weedy *B. campestris* (Brassicaceae): A risk of growing genetically modified oilseed rape. *American Journal of Botany* 81: 1620–26.
51 Atkinson, C (2000) Farmer puzzled by weed killer's surprise crop. *Sydney Morning Herald*, Saturday 5 February (reprinted from the *Guardian*).
52 Preston, C (2000) The use of herbicide tolerant crops. In *Weed Management 2000*. The Weed Society of NSW, Sydney, pp. 3.1–3.4.
53 Beard, BH (1981) The sunflower crop. *Scientific American* 244(5): 124–31.
54 Ashby, J (2000) Hussar. In *Weed Management 2000*. The Weed Society of NSW, Sydney, pp. 12.1–12.6.
55 Heisey, RM (1996) Identification of an allelopathic compound from *Ailanthus altissima* (Simaroubaceae) and characterization of its herbicidal activity. *American Journal of Botany* 83: 192–200.
56 Murray, DR (2000) Wind, water and weeds. In David R Murray *Successful Organic Gardening*. Kangaroo Press, Sydney, pp. 48–51.
57 Atkinson, T (2000) Flame weeding. In *Weed Management 2000*. The Weed Society of NSW, Sydney, pp. 6.1–6.7.
58 Mulligan, M & Hill, S (2001) Thinking like an ecosystem. In M Mulligan and S Hill *Ecological Pioneers: A Social History of Australian Ecological Thought and Action*. Cambridge University Press, Melbourne, pp. 191–215.
59 Medd, RW (2000) Some advances in weed management for cropping systems. In *Weed Management 2000*. The Weed Society of NSW, Sydney, pp. 2.1–2.5.
60 Carter, RJ (2000) Weed management strategies and plans — a global, national and state perspective. In *Weed Management 2000*. The Weed Society of NSW, Sydney, pp. 8.1–8.3.

4
SETTING PRIORITIES FOR PLANT IMPROVEMENT

> ... your father's lunge to be first up in the genetically modified gherkin market hit a snag, seeing he can only fit one gherkin in the silo, and that's if he takes the roof off.
>
> Patrick Cook[1]

MISGUIDED GOALS

Why do cartoonists love to make fun of genetically modified plants? Probably because the people promoting the fruits of this technology have often been their own worst enemies. Extravagant and premature claims abound, and there is a dearth of clear-sighted planning for practical objectives. Many stated goals are trivial or downright stupid, such as producing turf grasses that light up when trodden on, Christmas trees that provide their own illumination,[2] or pears that taste like apples and vice versa. Why bother, when hundreds of varieties of apples and pears are maintained and grown around the world? There is plenty of choice already and, for those with limited space, there are multiply-grafted 'fruit-salad' trees on dwarfing rootstock available from progressive nurseries.

The 'silly season' for genetic modifications seems to have begun in the 1920s, with one of the standard jibes made about Luther Burbank, namely that he 'crossed the eggplant and the milkweed to make an omelette plant'.[3] It resumed in the 1960s, with Dr Kimball Atwood's speculation about future possibilities couched as follows:

> We could, for example, produce an organism that combines the happy qualities of animals and plants, such as one with a large brain so that

it can indulge in philosophy and also a photosynthetic area on its back so that it would not have to eat. It is not inconceivable that there could be humanoids with chlorophyll under their skins so that they would look like the enormous green man on a can of peas.[4]

Such fanciful speculation overlooks the surface area relationship to light interception and photosynthetic productivity that any competent plant physiologist might have pointed out, then or now. The surface area of a human back is pathetically inadequate to supply the food requirements of the human organism. Clearly the human species is going to remain heavily dependent on plants for food.

What else is nonsense? Putting genes for Antarctic fish 'antifreeze' serum glycoproteins into fruits like tomatoes or strawberries to help them retain texture in cold storage comes into the 'downright stupid' category of potential modifications. These interesting glycoproteins contain repeating units of alanine-alanine-threonine, with each threonine linked to a disaccharide consisting of galactosyl-N-acetyl galactosamine.[5] The genes for several new enzymes would be necessary for this synthesis in plants, not just one, so the production of this glycoprotein in plants would be technically difficult to bring about.

A 1% solution of this glycoprotein is required to depress freezing point just 1°C below zero.[5] But so little glycoprotein would be produced in a fruit that the whole exercise is futile. Suppose the novel glycoprotein were synthesised to the extent that it represented 1% of all the protein in the fruit. If this were 2% of fruit fresh weight, then the glycoprotein would represent only 0.02% of fruit fresh weight, depressing freezing point about 0.02°C. This is scarcely perceptible, and not worth the expense of the attempt.

Other goals of genetic engineering are apparently benign, but still a massive waste of money compared with simpler options. Florigene's nebulous blue rose falls into this category. Rose growers concluded long ago that a blue rose was 'beyond reach',[6] and as noted elsewhere, 'the world does not lack blue flowers'.[7]

The most worthwhile breeding objectives have often been left for conventional breeders to worry about, or postponed in favour of misguided goals, such as generating plants resistant to herbicides (Chapter 3). This unfortunate priority resulted from the inherent conflict of interest that prevailed in burgeoning agrichemical companies as they evolved into vertically integrated agribusinesses. They are still evolving (Table 4.1), and a recent trend is for the separation of agribusiness from more profitable pharmaceutical activities. An exception is seen in the case of Du Pont and Protein Technologies International (PTI), where the medical benefits demonstrated in successful trials of PTI's soy protein isolate allow a marketing combination described as 'nutraceutical'.[8]

Table 4.1
Companies with the largest sales of agrichemical products in 1997[a]

Company	Sales (US$ million)
Aventis[b]	4554
Novartis[c,d]	4199
Monsanto[e]	3126
AstraZeneca[d]	2674
Du Pont[f]	2518
Bayer	2254
Dow Agricultural Sciences	2200
American Home Products[g]	2119
BASF	1855
Sumitomo	717

[a]Figures from the Rural Advancement Foundation International (RAFI) (1997).
[b]Encompassing AgrEvo, Rhone Poulenc and Hoechst.
[c]Novartis was formed from Sandoz and Ciba Geigy in 1996.
[d]Novartis and AstraZeneca combined their seed interests and agrichemical business to form Syngenta in October 2000.
[e]Now merged with Pharmacia & Upjohn.
[f]Acquired soybean processor Protein Technologies International (PTI) in 1998 and merged with Pioneer HiBred in 1999.
[g]Merged with Monsanto in 1998.

In this chapter and the next, many of the goals of those who intend to modify plants using recombinant nucleic acid techniques are described. Some aims are worthwhile, and success would increase yields by helping to prevent losses due to pests and diseases or by making plants more adept at acquiring nutrients from the soil. Nevertheless, it is foolhardy to assume that all of the currently available methods for genetic modification are equally safe, and it is unscientific to accept the new technology *carte blanche*. A critical appraisal of these breeding goals accompanies their description. A discussion of risk assessment follows in Chapter 6.

USING PLANT PROTEINS FOR DEFENCE

Two kinds of defensive plant proteins or glycoproteins are generally well represented in seeds: lectins and proteinase inhibitors. These proteins are active against some major pests of the seeds in which they occur, and they are clear candidates for built-in insecticides if they could be produced in the leaves, stems, flowers or young fruits as well as in the seeds.

Lectins are 'proteins or glycoproteins characterized by their ability to bind particular sugar residues that belong to polysaccharide moieties of glycoproteins, glycolipids, polysaccharides, or simple glycosides'.[9] In medical research, lectins are well known for their ability to clump red blood cells according to blood group, or to stimulate cell division (mitosis) in other kinds of cell by binding to the cell surface. Biochemically, they are extremely useful in the separation of glycoproteins by affinity chromatography.

Seed lectins and proteinase inhibitors are toxic to humans only when the source is eaten raw. They are normally denatured by cooking, for instance, by steaming bean seeds for 20 minutes. Lectins that are not denatured can bind selectively to carriers responsible for sugar uptake in the gastric mucosa, and some proportion of bound lectin can be ingested through the mucosa.[10] Proteinase inhibitors will not bind to the mucosa, but will bind to the digestive enzymes trypsin and/or chymotrypsin, unless they are first denatured by cooking.[11]

A major attraction in using the genes for such proteins from edible seeds is that there should be no unexpected toxic effects from gene expression, because these gene-products have already been part of our staple foods for at least 10 000 years. Obviously they should not be put into plants that are customarily eaten raw, such as lettuce. With this proviso, complications could arise only from other components of the constructs used to incorporate the desired genes into the genome, or from disruption of the functioning of normal genes by the insertion of the construct.

INSECTICIDAL PROTEINS FROM LEGUMES

Moving genes that are directly responsible for insecticidal proteins or glycoproteins is much simpler than transferring a suite of genes necessary for the synthesis of all the enzymes belonging to a complex metabolic pathway. One of the first proposals of this kind was to have the gene for a proteinase inhibitor from cowpea (*Vigna unguiculata*) seed expressed in the leaves of various other plants, so that it might act against the caterpillars of leaf-eating insects.[12] The cowpea trypsin/chymotrypsin inhibitor with best activity against larvae of the seed weevil *Callosobruchus maculatus* was shown to be active against a wide range of insect pests, including *Heliothis virescens* (tobacco budworm), *Heliothis zea* (corn earworm), *Helicoverpa armigera* (cotton bollworm), *Spodoptera littoralis* (army worm), *Manduca sexta* (tobacco hornworm) and *Locusta migratoria* (locust).[12,13,14]

Tobacco plants (*Nicotiana tabacum*) were transformed using an *Agrobacterium* vector, and these proved to be well protected against tobacco budworm compared with plants transformed with the proteinase inhibitor gene placed in the reverse orientation to its promoter sequence.[12,15]

More recently, the gene for an amylase inhibitor from bean seeds (*Phaseolus vulgaris*) has been added to the genome of a field pea (*Pisum sativum*), to protect stored seeds against damage from pea weevils (*Bruchus pisorum*). The bean cultivar Tendergreen was used as the source of the amylase inhibitor gene, and the transformation was performed using an *Agrobacterium* plasmid.[16] Field peas are generally round-seeded and grown for soup or for supplementary animal feeding. If all goes well with the field trials, this weevil-resistant pea could be released by CSIRO Plant Industry in 2004. It is noteworthy, and commendable, that an extra two years have been spent in the development of this pea to allow for the removal of a herbicide resistance gene used initially as a selectable marker.

SNOWDROP LECTIN AS AN INSECTICIDE OR NEMATICIDE

The lectin from snowdrop (*Galanthus nivalis*) is not one of those that has been eaten habitually by humans. It binds to the hexose sugar mannose, and was chosen for incorporation into a potato because of its efficacy against aphids[17,18] and nematode worms.[10] Because the common viral promoter CaMV35S was employed in the construct used to bring about the transformation, the snowdrop lectin is synthesised in all parts of the transgenic plant, including the tubers intended for human consumption. For this reason, animal feeding experiments were performed, with transformed potato, unmodified potato (Desiree), and unmodified potato to which purified snowdrop lectin had been added.[10] These experiments gave very interesting results, and the controversy over their interpretation has received wide publicity (see Chapter 6).

USING BT TOXIC PROTEINS AS INSECTICIDES

During spore formation, the soil bacterium *Bacillus thuringiensis* synthesises an array of crystalline proteins that are toxic to insect larvae because they break down cells in the gut wall, preventing food uptake. More than 40 different proteins have been found from various strains of this bacterium, and since 1989 they have been classified according to an arbitrary system that reflects how similar their amino acid sequences are.[19] The abbreviation 'Cry' is followed by roman numerals for the first distinction, then capital letters, then lower case letters in parentheses. Gene technologists have been trialling these proteins one at a time; for example, CryIA(b) has been expressed in tomato,[20] CryIA(c) in INGARD cotton produced by Monsanto, and CryIXC in StarLink maize produced by Aventis. There are wide differences in susceptibility shown by various pest insects across several orders; for example, *Helicoverpa virescens* is susceptible to CryIA(c) to about the same extent as *Manduca sexta* and *Plutella xylostella* (diamondback or cabbage moth), but *Heliothis zea* is 300 to 500-fold less susceptible, and *Spodoptera* species are not affected.[21] Cotton plants producing CryIA(b) are not as resistant to *Helicoverpa* species as those producing CryIA(c).[22] Expressing the gene for a single crystalline protein means that transgenic plants cannot be as potent as any strain of

Bacillus thuringiensis that produces its own complex mixture. There is a natural synergism at work[23] that cannot yet be replicated in a transgenic plant. The problems encountered with Bt-transformed crop plants are discussed in Chapter 6.

IMPROVING RESISTANCE TO VIRUSES

Plant viruses are transmitted by sap-sucking and leaf-chewing insects. Often insecticides are used on crop plants not because of insect pests *per se*, but because of the attendant risk of their spreading viruses. Virus-resistant varieties obviate the need for insecticides, and even partial resistance is valuable, as extreme symptoms can be avoided for the longest possible times.

Resistance genes can be acquired from varieties discovered to be resistant in screening tests. But conventional breeding to add virus resistance to susceptible cultivars becomes very complicated and time-consuming, as it involves so-called 'bridging genotypes' and frequent backcrossing (for example as in *Cucurbita*).[24,25] Nevertheless, this conventional approach is effective and demonstrably safe.

There are several ways to confer virus resistance via recombinant DNA (or RNA) technology. First, incorporating an 'antisense' sequence to a viral nucleic acid sequence has been tried. Since most plant viruses possess a core of RNA, the aim has been to incorporate a DNA sequence that codes for an RNA sequence that is complementary to part of a viral RNA gene sequence. Once transcribed, such RNA sequences should bind to the complementary parts of the infecting viral RNA, thus preventing normal expression in the host cells.

Although 'antisense' modifications have worked well for some purposes, their use with viral sequences has been disappointing. At best, the 'resistance is most often expressed as an increase in the proportion of plants escaping infection, but delayed symptom appearance and reduced severity are also observed'.[26] Another assessment of this approach is quite uncomplimentary: 'Although some protection has been observed, it has been quite weak and appears to have little commercial potential in its present state of development.'[27]

A different strategy for modifying a plant's viral resistance involves the use of synthetic 'satellite' RNA sequences. Some viruses contain satellite sequences that can ameliorate the symptoms of infection by that virus. Alternatively, these sequences can make the symptoms more severe. DNA complementary to the desired satellite transcript is incorporated into the plant genome. On infection with the virus, the synthesis of 'satellite' RNA apparently competes with the synthesis of core viral RNA, thus reducing the rate of assembly of new viruses. However, there are appreciable hazards inherent in this approach. Gail Timmerman[26] sets out three main reasons for caution:

- Satellite RNAs that are protective in one species might endanger other crop species;
- Satellite RNAs can mutate rapidly, and a single base change can be enough to alter a beneficial sequence to one that enhances virulence;
- Recombination can occur between different satellite RNA sequences.

The need for caution is endorsed by Sorenson, Provvidenti and Munger:[27] 'Obviously the risks inherent in such an approach preclude its usefulness until we have a better understanding of the dynamics of satellites and their role in viral etiology.' Timmerman is nevertheless optimistic, suggesting that: 'Once there is a more advanced understanding of the molecular biology of satellite RNAs, it may be possible to manipulate the sequences of these agents so that they lose their transmissibility from the transgenic plant and yet retain their efficacy in ameliorating plant viral disease.'[26]

As an unexpected benefit from their studies of satellite RNA, Wayne Gerlach and Jim Haseloff were able to use its ability to cut RNA strands to design 'ribozymes'.[28] These are pieces of RNA that can cleave part(s) of the same molecule, or other RNA strands. This development became known as 'gene shears' technology, and was patented by CSIRO Plant Industry.

By far the most effective new method of obtaining virus resistance involves inserting and expressing some part or parts of the viral nucleic acid. Initially the emphasis has been on sequences that code for viral coat proteins.[26,27] The viral genome is very small, and these genes are readily accessible. The resulting synthesis of viral coat protein in plant cells rarely represents more than 0.1% of total protein, and yet the presence of the coat protein gene and its product can have a remarkable protective effect. Often this extends to closely related viruses that might possess similar, but not identical, coat proteins.

Adopting this approach, a transgenic tomato has been produced that shows partial resistance to *Physalis* mottle tymovirus.[29] This is an interesting case, because the virus assembles a coat consisting of 180 copies of a single protein. CSIRO Plant Industry scientists have developed white clover (*Trifolium repens*) resistant to alfalfa mosaic virus.[16] Subject to satisfactory feeding trials, the first release is expected in 2004. This virus-resistant clover is intended to benefit pasture growth and animal production.

Surprisingly, even better results have been obtained when a plant is modified to express non-transcribed regions of viral RNA. When RNA transcripts generated by an invading virus are being synthesised, the extra pieces produced by the transgenes form regions of double-stranded RNA, which form kinks. These kinks are extremely susceptible to breakdown by plant ribonucleases. Without new core RNA, viruses cannot be assembled. CSIRO Plant Industry scientists have pioneered the methodology of this approach, and hold patents that they can license in a reciprocal manner for other techniques that they need. This technique has already been used to confer immunity to barley yellow virus on wheat.

IMPROVING RESISTANCE TO FUNGI

Plants can block fungal invasion by a combination of preformed passive barriers, generally compounds present in cell walls, and 'pathogen induced active responses'.[30] To date, however, there has been little recourse to recombinant DNA techniques in breeding programs aimed at improving fungal resistance. Major fungal pathogens of cotton and canola have been countered by conventional breeding techniques, for example *Fusarium* resistance in cotton and 'blackleg' resistance in canola (Pacific Seeds' variety Surpass-400). Resistance to various rust diseases of wheat has been and remains a major priority for conventional wheat breeding in Australia.[31]

Antisense techniques appear to offer some promise once an infection-responsive protein has been identified. The agent that causes clubroot in the mustard family (Brassicaceae or Cruciferae) is not quite a fungus, but a close ally — this is the 'slime mould' *Plasmodiophora brassicae*. The symptoms of this disease are delayed when *Arabidopsis thaliana* plants are transformed with genes for the enzyme nitrilase in an antisense orientation.[32]

IMPROVING RESISTANCE TO BACTERIA

Natural resistance to many bacterial diseases of plants appears to be complex or polygenic in its inheritance, and is often ineffective at higher growing temperatures.[33] Nevertheless, opportunities have been taken to confer resistance to various species of *Erwinia* by using *Agrobacterium*.[34,35,36] Potatoes have been made resistant to *Erwinia carotovora*,[34] and some apple cultivars have been made resistant to fireblight, *Erwinia amylovora*.[36] This latter disease is spread by pollinating insects and is very difficult to control. It would inevitably spread further if apples from New Zealand were permitted to enter Australia, which is currently free of fireblight. Quarantine should clearly be maintained as the first line of defence. An alert quarantine system is much cheaper than conferring resistance on every single apple cultivar, which is the only realistic approach to eradication for countries that harbour the disease.

In California, a silkworm (*Bombyx mori*) gene has been transferred to a grape cultivar, Thompson's Seedless, to confer resistance to Pierce's disease, which affects the vascular system, specifically the xylem cells.[37] Grapevines that produce the silkworm protein cecropin are able to resist infection by the bacterium responsible for Pierce's disease. Cecropin binds selectively to the membranes of bacterial cells and punctures them. One difficulty with this modification stems from using a viral promoter, which means a lack of organ specificity in the expression of the gene. At this stage the production of cecropin is not confined to the xylem cells of the main stems and stalks — this protein will also appear in the fruit in very small amounts.

MODIFYING PHYSIOLOGICAL PARAMETERS
IMPROVING ALUMINIUM TOLERANCE
In acid soils, positively charged forms of aluminium such as Al^{3+} inhibit plant growth by binding to the surfaces of root cells. The ability to avoid aluminium toxicity depends on the capacity of the root system to secrete organic acid anions, such as malate and citrate. This was discovered first for wheat, with some varieties showing ten times the capacity for malate secretion relative to cultivars that are sensitive to aluminium.[38] More recently, white lupin (*Lupinus albus*) has been shown to secrete copious amounts of citrate from the roots.[39]

This ability can be conferred on other legumes, such as the pasture species lucerne (*Medicago sativa*) and subterranean clover (*Trifolium subterraneum*), by transferring genes for the enzyme citrate synthase from suitable bacteria. The bacterial enzyme differs in its regulation from the endogenous plant enzyme, and can make more citrate available for secretion from the root surfaces. A bacterial enzyme transferred to tobacco (*Nicotiana tabacum*) increased the rate of secretion of citrate from the roots by a factor of three.[40]

INCREASING PHOSPHORUS UPTAKE
Roots take up inorganic phosphate, which is present in soil only in very low concentrations (micromolar). Many plant species have roots that are capable of associating with mycorrhizal fungi, which facilitate their uptake of phosphorus and other minerals. Nevertheless, phosphorus-rich fertilisers need to be added to agricultural soils regularly, at least once every two years. This requirement is a major concern, as readily available sources of phosphatic fertilisers are dwindling, and they are becoming more expensive.

However, agricultural soils contain a very large reserve of phosphorus that is unavailable to plants. This has accumulated from the times when annual applications of superphosphate were routinely given, whether needed or not.[41] Some of this unavailable phosphorus is inorganic, such as the insoluble compound calcium phosphate, but much is organic. It has been estimated that of every 10 kilograms of phosphorus in new fertiliser applied per hectare, only 10 to 20 per cent is available to plants in the year of application.[42] More of this phosphorus is available over three years, but about half will contribute to the unavailable reservoir of soil phosphorus. Many soils that have been cultivated continuously contain unavailable phosphorus equivalent to more than ten times an annual application of fertiliser.

It would be extremely useful if plants could be modified to use more of this phosphorus. Much less fertiliser would then be needed. One approach is to enhance organic acid secretion from the roots, as for plants with improved aluminium tolerance. A second approach involves adding the gene for a specific phosphatase, and altering this enzyme's 'leader' sequence to ensure that it is secreted by roots, rather than being concentrated internally.[43] This would allow the roots of the modified plant to liberate phosphate from soil phytate (inositol hexa-phosphate, similar to the material stored in seeds, see below). Phytase enzymes from soil micro-organisms, such as the fungus *Aspergillus niger*[43] or species of the soil bacterium *Pseudomonas*,[44] are excellent candidates for transfer. Subterranean clover modified to produce a secreted phytase has been developed by CSIRO Plant Industry to the point where it is being field tested.

Using genetically modified plants is not the only solution to the phosphorus problem. Culturing the right kinds of soil micro-organisms, and adding these to soils with high reservoirs of unavailable phosphorus, is also worthwhile. With regard to such cultures, however, 'their widespread application remains limited by a poor understanding of microbial ecology and population dynamics in soil and by inconsistent performance over a range of environments.'[45]

REDUCED BROWNING IN FRUITS

A proposal to reduce the brown coloration of dried grapes by blocking the enzyme polyphenol oxidase (PPO) is being pursued by CSIRO Horticulture. This has been accomplished by gene silencing — incorporating an antisense sequence to the gene for this enzyme. There are two immediate criticisms of the proposal. First, so far as the consumer is concerned, low colour is correlated with lower quality. A pale raisin or currant simply would not sell, as the consumer expects more intense flavour to accompany deeper brown coloration. Sales of dried grape products are in the doldrums, and something like this would be certain to depress them further. Second, a blanket inhibition of this enzyme in the developing grape berry might have deleterious consequences, especially with regard to loss of pest or disease resistance. Phenolic compounds are extremely important in natural defence systems. Removing PPO is not a goal that any intelligent plant breeder should be entertaining.

The loss of functional PPO activity has already occurred in Goldfinger banana. The skin of this banana does not brown or blacken, but remains yellow. Although welcomed by the industry, this modification conceals bruising, which normally accompanies the handling and transportation of bananas to markets and shops. The discerning consumer thus loses a clue to quality, and might refuse to purchase bananas altogether rather than persevere with bruised ones that look all right. This kind of modification is of no practical benefit to either producer or purchaser. With domestic consumption averaging only 15 kilograms per head per year, banana producers would be very brave to insist on growing varieties like this one.

TERMINATOR

'Terminator' was the name given by Rural Advancement Foundation International (RAFI) to proposals involving the expression of a bacterial ribonuclease, barnase, in seeds. This is intended to provide a built-in method for dissuading farmers from saving and replanting seeds produced by plants derived from the seeds originally purchased. For the seed company, this is better than selling F_1 hybrids, whose progeny do not breed uniformly, but are at least capable of growing and reaching maturity. There is a great deal of confusion as to how Terminator is supposed to work. But it does not produce seeds that are already dead, as claimed by Martha Crouch,[46] nor are the seeds sterile, as described by Mae Wan-Ho and colleagues.[47] In this context the term is misapplied, because if the plants being grown were sterile, there would be no seeds at all.

As described in Chapter 2, Terminator involves the ribonuclease from *Bacillus amyloliquefaciens* called barnase, and its natural inhibitor, bar or barstar, from the same source. While barstar is bound to barnase, this enzyme remains inactive, but unbound barnase is able to destroy RNA. Seeds are modified to express barnase by responding to a pretreatment provided by the seed company, such as the antibiotic tetracycline, which prompts an enzyme to remove a blocking sequence from the barnase promoter. The use of the antibiotic tetracycline as a gene switch is now widespread.[48]

How is it possible for seed tissues to synthesise barnase without disrupting the normal processes of seed development and maturation? As seeds develop, protein synthesis accelerates and reserve proteins are manufactured, as well as structural proteins and enzymes. Different proteins have characteristic times of appearance and distinct periods when their maximum rates of synthesis are observed.[9] So to place barnase in an almost mature seed without significantly affecting the composition or weight of the seed, a promoter has to be attached to the barnase gene from the gene for a seed protein that is normally manufactured very late in seed development.

The seeds of all major crop species dry out as they mature, reaching a stable low water content around 6–11 per cent of seed weight. This value varies with species and variety. Protein synthesis continues during the period of desiccation and shrinking, but eventually metabolic activity practically ceases. Everything necessary for the resumption of protein synthesis is stored, including ribosomes, transfer RNA, and even some messenger RNA.[49] When water is taken up again by the seed, cell expansion allows the embryonic root to extend and penetrate the seedcoats, thus bringing about germination. Cell division resumes, and although some stored proteins are being broken down,[50] new proteins are also being synthesised. This new protein synthesis can be detected as early as one hour after water uptake has commenced.[49]

A ribonuclease such as barnase, acting in the absence of barstar, would soon begin to destroy stored RNA of all kinds. This would not prevent germination, as cell expansion alone is sufficient to enable the radicle to break the seedcoat. But it would certainly bring seedling growth to a halt. Although selective breakdown of stored messenger RNA normally accompanies germination and seedling establishment, barnase would break down newly synthesised messenger RNA. Barnase would also destroy ribosomal RNA, removing the sites of protein synthesis. Without new protein synthesis, there can be no new growth. So the expression of barnase can be likened to a lethal mutation, such as those that prevent the synthesis of chlorophyll. The seeds are not dead, but the process of seedling establishment is unable to proceed beyond a certain point.

When it appeared that Monsanto were about to take over Delta and Pine, who in conjunction with the United States Department of Agriculture were first to develop this application of Terminator to seeds, Robert Shapiro stated that Monsanto would not use the technology.[51] But the takeover did not proceed, and Delta and Pine have not ruled out using this system in the future. Other companies also have their versions of Terminator, and the merger of Novartis with AstraZeneca (Table 4.1) leaves many of the most recent patents in this area in the hands of Syngenta.[52] This is a most unfortunate situation, and one that will require continued surveillance and over-riding legislation.

SUMMARY

As indicated at the beginning of this chapter, improving resistance to pests and diseases is a fruitful area for genetic modification by the new methodologies, and most of the examples given here illustrate reasonable goals. Modifying physiological parameters is more problematic, but again some significant and worthwhile goals have been advanced. How safely these goals might be attained is discussed in Chapter 6, after consideration of utilitarian breeding goals (Chapter 5).

REFERENCES

1   Cook, P (2000) Man discovers outback. *The Bulletin* 15 February, p. 114.
2   von Radowitz, J (1999) Pine for a little yuletide glow. *The Australian* 27 October.

3   Hall, W (1927) Luther Burbank — naturalist. In Luther Burbank and Wilbur Hall *The Harvest of the Years*. Houghton Mifflin, Boston and New York, pp. ix–xxvi.
4   Atwood, K (1965) Discussion — Part 1. In TM Sonneborn (ed.) *The Control of Human Heredity and Evolution*. Macmillan, New York, pp. 35–38.
5   Ahmed, AI, Osuga, DT & Feeney, RE (1973) Antifreeze glycoprotein from an Antarctic fish — Effects of chemical modifications of carbohydrate residues on antifreeze and antilectin activities. *Journal of Biological Chemistry* 248: 8524–27.
6   Pietsch, EB (ed.) (1975) *The Australian Rose*. The Horticultural Press, Melbourne.
7   Murray, DR (1999) Breeding new peas and beans. In David R Murray *Growing Peas and Beans*. Kangaroo Press, Sydney, pp. 55–58.
8   Stipp, D (1998) Engineering the future of food. *Fortune* 28 September, pp. 128, 130, 134, 136, 140, 144.
9   Murray, DR (1984) Accumulation of seed reserves of nitrogen. In David R Murray (ed.) *Seed Physiology Volume 1. Development*. Academic Press, Sydney, pp. 83–137.
10 Ewen, SWB & Pusztai, A (1999) Effects of diets containing genetically modified potatoes expressing *Galanthus nivalis* lectin on rat small intestine. *The Lancet* 354: 1353–54.
11 Belitz, H-D & Weder, JKP (1990) Protein inhibitors of hydrolases in plant foodstuffs. *Food Reviews International* 6(2): 151–211.
12 Gatehouse, JA (1991) Breeding for resistance to insects. In David R Murray (ed.) *Advanced Methods in Plant Breeding and Biotechnology*. CAB International, Oxford, pp. 250–76.
13 Gatehouse, AMR, Gatehouse, JA, Dobie, P, Kilminster, AM & Boulter, D (1979) Biochemical basis of insect resistance in *Vigna unguiculata*. *Journal of the Science of Food and Agriculture* 30: 949–58.
14 Gatehouse, AMR, Gatehouse, JA & Boulter, D (1980) Isolation and characterisation of trypsin inhibitors from cowpea (*Vigna unguiculata*). *Phytochemistry* 19: 751–56.
15 Hilder, VA, Gatehouse, AMR, Sheerman, SE, Barker, RF & Boulter, D (1987) A novel mechanism of insect resistance engineered into tobacco. *Nature* 330: 160–63.
16 Higgins, TJV (Deputy Director, CSIRO Plant Industry) (2001) personal communication, 3 May.
17 Down, RE, Gatehouse, AMR, Hamilton, WDO & Gatehouse, J (1996) Snowdrop lectin inhibits development and decreases fecundity of the glasshouse potato aphid (*Aulacorthum solani*) when administered in vitro and via transgenic plants both in laboratory and glasshouse trials. *Journal of Insect Physiology* 42: 1035–45.
18 Gatehouse, AMR, Down, RE & Powell, KS (1996) Transgenic potato plants with enhanced resistance to the peach-potato aphid *Myzus persicae*. *Entomology Experimental and Applied* 79: 295–307.
19 Hofte, H & Whiteley, HR (1989) Insecticidal crystal proteins of *Bacillus thuringiensis*. *Microbiological Reviews* 53: 242–55.
20 Noteborn, HPJM, Bienenmann-Ploum, ME & van den Berg, JHJ (1995) Safety assessment of the *Bacillus thuringiensis* Insecticidal crystal protein CryIA(b) expressed in transgenic tomatoes. In K-H Engel, GR Takeoka and R Teranishi (eds) *Genetically Modified Foods: Safety Issues*. ACS Symposium Series 65, Washington DC, pp. 134–47.
21 Chilcott, CN & Wigley, PJ (1994) Insecticidal activity of *Bacillus thuringiensis* crystal proteins. In RJ Ackhurst (ed.) *Proceedings of the 2nd Canberra Meeting on Bacillus thuringiensis*. CSIRO Division of Entomology, Canberra, pp. 43–52.

22 Llewellyn, D, Last, D, Mathews, A, Hartweck, L, Fitt, G, Peacock, WJ & Buehler, R (1994) Transgenic Australian cotton cultivars expressing insecticidal protein genes from *Bacillus thuringiensis*. In RJ Ackhurst (ed.) *Proceedings of the 2nd Canberra Meeting on Bacillus thuringiensis*. CSIRO Division of Entomology, Canberra, pp. 69–73.
23 Frederici, BA & Wu, D (1994). Synergism of insecticidal activity in *Bacillus thuringiensis*. In RJ Ackhurst (ed.) *Proceedings of the 2nd Canberra Meeting on Bacillus thuringiensis*. CSIRO Division of Entomology, Canberra, pp. 23–30.
24 Kyle, MM (ed.) (1993) *Resistance to Viral Diseases of Vegetables: Genetics and Breeding*. Timber Press, Portland, Oregon.
25 Herrington, ME, Loader, L, Prytz, S & Slade, A (1999) Incorporating etch and virus resistance into butternut pumpkins. *Proceedings 11th Australian Plant Breeding Conference*, CRC for Molecular Plant Breeding, Adelaide, Volume 2: 58–59.
26 Timmermann, GM (1991) Genetic engineering for resistance to viruses. In David R Murray (ed.) *Advanced Methods in Plant Breeding and Biotechnology*. CAB International, Oxford, pp. 319–39.
27 Sorenson, JC, Provvidenti, R & Munger, HM (1993) Conclusions: Future prospects, strategies and problems. In Molly M Kyle (ed.) *Resistance to Viral Diseases of Vegetables: Genetics and Breeding*. Timber Press, Portland, Oregon, pp. 255–64.
28 Buck, K (1989) Brave new botany. *New Scientist* 122(No. 1667): 32–35.
29 Sree Vidya, CS, Manoharan, M, Ranjit Kumar, CT, Savithri, HS & Lakshmi, SG (2000) *Agrobacterium*-mediated transformation of tomato (*Lycopersicon esculentum* var. Pusa Ruby) with coat-protein gene of *Physalis* mottle tymovirus. *Journal of Plant Physiology* 156: 106–10.
30 Chakravorty, AK & Scott, KJ (1991) Resistance to fungal diseases. In David R Murray (ed.) *Advanced Methods in Plant Breeding and Biotechnology*. CAB International, Oxford, pp. 277–98.
31 McIntosh, RA, Wellings, CR & Park, RF (1995) *Wheat Rusts — An Atlas of Resistance Genes*. CSIRO Australia, Melbourne.
32 Neuhaus, K, Grsic-Rausch, S, Sauertaig, S & Ludwig-Muller, J (2000) *Arabidopsis* plants transformed with nitrilase 1 or 2 in an antisense direction are delayed in clubroot development. *Journal of Plant Physiology* 156: 756–61.
33 Whalen, M (1991) Advances in breeding for resistance to bacterial pathogens. In David R Murray (ed.) *Advanced Methods in Plant Breeding and Biotechnology*. CAB International, Oxford, pp. 299–318.
34 During, K, Porsch, P, Fladung, M & Lorz, H (1993) Transgenic potato plants resistant to the phytopathogenic bacterium *Erwinia carotovora*. *Plant Journal* 3: 587–98.
35 Mourgues, F., Brisset, M-N & Chevreau, E (1998) Strategies to improve plant resistance to bacterial diseases through genetic engineering. *Tibtech* 16: 203–10.
36 Norelli, JL & Aldwinckle, HS (2000) Transgenic varieties and rootstocks resistant to fireblight. In Joel L Vanneste (ed.) *Fire Blight — the Disease and its Causal Agent, Erwinia amylovora*. CAB International, Oxford, pp. 275–92.
37 Samuel, E (2001) Vin de silkworm. *New Scientist* 170(No. 2292): 6.
38 Ryan, PR, Delhaize, E & Randall, PJ (1995) Characterisation of Al stimulated efflux of malate from the apices of Al tolerant wheat roots. *Planta* 196: 103–11.
39 Keerthisinghe, G, Hocking, PJ, Ryan, PR & Delhaize, E (1998) Effect of phosphorus supply on the formation and functions of proteoid roots of white lupin (*Lupinus albus* L.). *Plant, Cell and Environment* 21: 467–78.
40 de la Fuente, JM, Ramirez-Rodriguez, V, Cabrera-Ponce, JL & Herera-Estrella, L (1997) Aluminium tolerance in transgenic plants by alteration of citrate synthesis. *Science* 276: 1566–68.

41 Davidson, BR & Davidson, HF (1993) The sub — super revolution. In BR Davidson and HF Davidson *Legumes — The Australian Experience*. Research Studies Press, Taunton, UK, pp. 246–94.
42 Holford, ICR (1997) Soil phosphorus, its measurements and its uptake by plants. *Australian Journal of Soil Research* 35: 227–39.
43 Richardson, AE, Hadobas, PA & Hayes, JE (2001) Extracellular secretion of *Aspergillus* phytase from *Arabidopsis* roots enables plants to obtain phosphorus from phytate. *Plant Journal* 25: 641–49.
44 Richardson, AE & Hadobas, PA (1997) Soil isolates of *Pseudomonas* spp. that utilize inositol phosphates. *Canadian Journal of Microbiology* 43: 509–16.
45 Richardson, AE (2001) Prospects for using soil microorganisms to improve the acquisition of phosphorus by plants. *Australian Journal of Plant Physiology* 28: 897–906.
46 Crouch, ML (1998) How the Terminator terminates. <http://www.biotech-info.net/howto.html>, accessed 2000.
47 Ho, M-W, Cummins, J & Bartlett, J (2000) The killing fields — Terminator crops at large. Institute of Science in Society (ISIS), <http://www.i-sis.org/terminatorstory-pr.shtml>, accessed 2000.
48 Spinney, L (2000) Switched on. *New Scientist* 168(No. 2263): 52–55.
49 Simon, EW (1984) Early events in germination. In David R Murray (ed.) *Seed Physiology Volume 2. Germination and Reserve Mobilization*. Academic Press, Sydney, pp. 77–115.
50 Murray, DR (1984) Axis-cotyledon relationships during reserve mobilization. In David R Murray (ed.) *Seed Physiology Volume 2. Germination and Reserve Mobilization*. Academic Press, Sydney, pp. 247–80.
51 Specter, M (2000) The food that bit back. *Good Weekend*, the *Sydney Morning Herald* Magazine, 10 June, pp. 18–24.
52 Warwick, H (2000) Syngenta — switching off farmers' rights? GeneWatch UK, <http://www.genewatch.org>, accessed 2001.

5
PROPOSALS WITH NUTRITIONAL, MEDICAL OR UTILITARIAN GOALS

> Those who propose changes carry a heavy burden of proof that would justify them.
> Lord Dainton[1]

ALTERING NUTRIENT COMPOSITION

Just as the first releases of genetically modified plants are supposed to benefit growers, the 'second generation' of genetically modified plants is supposed to be offering greater benefits to consumers through improvements in nutrient content. Are such changes really necessary? And must they depend on the new technology rather than the old?

RICE WITH ß-CAROTENE

Rice (*Oryza sativa*) is a very important cereal, often ranked second after wheat in estimated global annual production. Grown mainly in the tropics, but extending well into subtropical and temperate regions, rice is the staple cereal of about half the human population (see Chapter 8). Adding ß-carotene to rice has been justified as providing vitamin A to people who are starving: 'More than one million children are believed to die every year as a result of vitamin A deficiency, a toll that engineered rice could reduce dramatically.'[2] The number of such premature deaths is put at two million by Gordon Conway, President of the Rockefeller Foundation.[3]

Transferring a whole biosynthetic pathway is a splendid technical achievement, albeit an expensive one. This was done by a group working in Switzerland.[4] An *Agrobacterium* plasmid was used, with the inserted sequences including genes from the bacterium *Erwinia uredovora* and the daffodil (*Narcissus* sp.) (Table 5.1). A promoter from the rice storage protein glutelin was attached to the daffodil genes, and the ubiquitous cauliflower mosaic virus promoter (CaMV35S) was attached to the bacterial genes.

Table 5.1
Sources of synthetic pathway components used for transgenic 'golden' rice, patent numbers, and patent holders[4]

Step or process	Patent number	Patent owner
Agrobacterium transformation system	WO8603776 (1986)	Plant Genetic Systems (Aventis)
phytoene synthase gene from daffodil	applied for	University of Freiburg
phytoene desaturase gene from Erwinia uredovora	EP0393690 (1990)	Kirin Brewery
lycopene cyclase gene from daffodil	applied for	University of Freiburg
constructs including a carotenoid synthesis gene	WO9806862 (1998)	Calgene (Monsanto)
glutelin promoter (Gt 1) from rice endosperm	J6391085 (1988)	Noriinsho
CaMV35S promoter	US5106739 (1992)	Calgene (Monsanto)
acid phosphatase gene(reporter)	US5668298 (1997)	Eli Lilly

With offers from Monsanto and other patent holders to forego payments for use of their patented procedures, there are now plans to give the rice free to growers, in the greatest public relations stunt of all time. Apart from the hail of self-congratulatory publicity, moral arguments have been raised in attempts to silence critics. A spokesperson from Nestlé attending the World Economic Forum in Melbourne, Michael Garrett, stated that people in well-off nations have no right to deprive the Third World of benefits such as this rice represents.[5] This is a converse kind of logic as, given the often-justified criticisms of some of its own marketing activities in the Third World, Nestlé is hardly in a position to be critical of such people. The proposed gift of this golden rice sits incongruously with attempts by other Americans to patent basmati rice from India and jasmine rice from Thailand. The benefits that these varieties already bring to India and Thailand would be diminished by American rice growers and exporters if they succeed in undermining niche markets currently supplied by these nations.

The 'help the starving poor' justification for ß-carotene-producing rice results from an extraordinarily narrow focus. People who are starving have multiple deficiencies. Because protein intake is deficient, two proteins involved in the transport of vitamin A cannot be produced in the liver in adequate quantities.[6] So starving people cannot even use the stores of vitamin A as retinyl esters that they already have in their livers, let alone take up any more vitamin A or its precursors from dietary sources. Under these dire circumstances, injections are the only way to target retinol shortage, and ß-carotene in the diet, from whatever source, is not immediately useful.

The real solution to Third World starvation requires environmental repair, equitable land tenure, and the encouragement of ecologically sustainable systems of agriculture and horticulture (Chapter 8). Often these were present before the imposition of high-input Green Revolution agriculture. The aim should be to ensure that people under marginal or unpredictable circumstances everywhere have access to a balanced diet. This does not mean every nutrient packed into a single food. Rice itself has a low protein content, approximately six per cent of dry matter,[7,8] or less when polished. The lipid content is also very low, no more than two per cent,[9] which is too low to assist much uptake of lipid-soluble (hydrophobic) substances such as ß-carotene. Rice could never be sole source of energy and nutrients, with or without added ß-carotene.

Apart from the question of sufficient energy content (calories), enough variety should be introduced to the diet to provide every nutritional need. This principle of complementarity of food sources is ancient[6,10] and its validity has been confirmed and reinforced by modern biochemical analyses.[8,11] Better sources of ß-carotene in the tropics include mangoes, papaya (paw paw), sweet potatoes, bananas of the right kinds, and even carrots, as in Vietnam (Chapter 8). How many of those advocating ß-carotene rice as a panacea for Third World hunger would accept a single food on their own plate, every meal, every day?

REDUCING PHYTATE-PHOSPHORUS IN WHEAT

The compound phytate is derived from a cyclic hexose sugar (*myo-inositol*) with all six hydroxyl (–OH) positions substituted with phosphate groups. These six phosphate groups are negatively charged and normally associated with positively charged mineral ions (cations), especially potassium, magnesium, zinc, calcium and iron, to form phytin. Phytin is localised as deposits inside the protein storage bodies of seeds[12] and represents a major store of carbohydrate, phosphorus and minerals.

A proposal to 'improve' wheat grain by lowering the phytate content is puzzling, and apparently based on the misconception that phytate acts during digestion to deprive us of mineral ions. This is rarely true, although it can happen when some of the phosphate groups are not fully associated with cations in the first place, or are not released as phosphate by hydrolysis during digestion. Lowering the phytate content would lower not only the seed's content of phosphorus, but also the content of its associated minerals. Nutritionally, this would be a retrograde step. It would also slow seedling growth rate and hinder establishment, making the attempt to grow a crop from such impoverished seeds more difficult than with seeds from unmodified cultivars.

INCREASING THE IRON CONTENT OF RICE

Displacing potassium or other mineral cations from phytin with extra iron would require manipulation of transport proteins in membranes of several kinds, and probably an increase in the storage cells' phytate content. These modifications could be much more difficult than anticipated, as rice grains have a relatively low content of the protein bodies that house phytin deposits in the starch-rich cells of the endosperm.[7] Keeping in mind the comments made earlier about not packing our total nutritional requirements into a solitary food, it would be simpler and cheaper to grow legumes in addition to rice, as legumes have a much higher natural iron content than rice. Even eating unpolished rice instead of polished white rice would immediately double the apparent availability of iron, although unpolished rice would tend to go rancid faster.

RAISING THE VITAMIN E CONTENT OF CANOLA

There is very little published information on the vitamin E contents of common foods. However, certain oilseeds are recognised as superior sources. The oils from sesame seeds (*Sesamum indicum*) and maize (*Zea mays*) have higher vitamin E content than oil from groundnut (peanut; *Arachis hypogaea*).[13] Canola seed oil must have less vitamin E than sesame or maize oils, otherwise a proposal to increase the vitamin E content of canola would not have been made.

Canola is a relatively new kind of oilseed rape, with a much reduced content of a toxic fatty acid, erucic acid.[14,15] Canola, however, is a poor choice for modification to increase vitamin E. Even though there is no longer a problem with erucic acid content, glucosinolates, fibre and tannins all represent a liability because, for poultry, pigs and ruminant animals, 'the thyroid and liver are significantly affected'[14] by these other constituents of canola meal. Moreover, in poultry, 'leg problems can occur when rapeseed meal is fed'.[14] Rather than modify canola plants to improve their vitamin E content, it would be better to encourage farmers to grow alternative oil crops, where both the oil and the meal after pressing represent safer products.

MODIFYING THE FATTY ACID COMPOSITION OF SEED OILS

Plant oils are compounds of fatty acids with glycerol. Each of the three hydroxyl groups of glycerol can be linked to a fatty acid, usually the kinds listed in Table 5.2. Oils have a higher energy content per unit volume than starch, and represent a compact alternative storage material, especially in seeds.

Each fatty acid has a long chain of carbon atoms, which is hydrophobic (water-repelling). When no double bonds are present, the fatty acids are 'saturated'. If a single double bond is present, then the fatty acids are 'monounsaturated'. If two or more double bonds are present, then the fatty acids are 'polyunsaturated'. Each double bond creates a kink in the chain. Mammals lack the enzymes to introduce double bonds, and accordingly linoleic and linolenic acids are regarded as essential in the human diet.[16]

Table 5.2
The fatty acids most commonly present in seed oils

Name	Number of C atoms	Number of double bonds
palmitic	16	none
stearic	18	none
oleic	18	one
linoleic	18	two
linolenic	18	three
eicosenoic	20	one
erucic	22	one

Canola oil is low in saturated fatty acids, high in oleic acid (about 60 per cent) and very high in polyunsaturated fatty acids (about 30 per cent).[15] Peanut oil has a similar profile. For maximum nutritional benefits, oils high in polyunsaturated fatty acids are best consumed as the whole food from which they come, and in view of the foregoing discussion, peanuts are preferable to rape seeds. For cooking, however, oils rich in oleic acid but with low polyunsaturated content are preferable, to minimise the production of damaging free radicals. Olive oil is ideal, with about 10 to 14 per cent saturated fatty acids, 73 to 77 per cent oleic acid, and 7 to 8 per cent polyunsaturated fatty acids. The oil from macadamia nuts is close to this in composition, but more difficult to process because of the very hard seedcoat.

When oils are treated to produce margarine spreads that mimic butter, some of the unsaturated fatty acids change their isomeric form from *cis* to *trans* around each double bond. *Trans* fatty acids are undesirable because they are treated in the body like saturated fatty acids, and have an even more pronounced effect on cholesterol concentration. Of the common saturated fatty acids, stearic is preferable to palmitic, which is believed to have a greater cholesterol-elevating influence. Unfortunately, palm oil, rich in palmitic acid, is inexpensive, and so is used in many fish shops and other fast-food outlets.

Conventional genetics has already discovered a number of Mendelian genes that can have major effects on the fatty acid composition of plant oils. High linoleic/low oleic versus low linoleic/high oleic are simple alternatives.[17] Two copies of a recessive allele at a different locus confer a high ratio of stearic acid to palmitic acid. Some major shifts in composition have been effected for the Linola variety of linseed (*Linum usitatissimum*), which has been made to resemble sunflower (*Helianthus annuus*). However, all of these changes apply to lipids throughout the plant, not just to the seed oils, so the properties of cell membranes are altered, sometimes adversely.

By using seed-specific promoters, gene technologists can have desirable changes expressed in the seed oils only, without changing membrane composition in the rest of the plant and possibly upsetting its adaptation to temperature and light conditions. Dr Allan Green and his colleagues at CSIRO Plant Industry have also applied gene silencing,[18] in which transcription of the target gene is prevented by the formation of double-stranded or kinked RNA, which is more susceptible to defensive plant ribonuclease activity (as for the viral resistance procedures outlined in Chapter 4). Different varieties of cotton have been produced with an increase in stearic acid content from 2 to 40 per cent in one, and an increase in oleic acid content from 20 to 75 per cent in another. These modified varieties are three to five years from release, and would be useful for cooking oil and margarine production, helping to reduce the occurrence of unsaturated fatty acids with the *trans* configuration. A combination of these two types of cotton is envisaged that would produce a material with the properties of cocoa butter.

INCREASING SEED PROTEIN CONTENT

It is possible to increase the amount of protein in seeds of a given species, but two considerations help to place such proposals in context. First, the most widely used method of estimating seed protein content is inaccurate. This is the simple procedure of measuring the nitrogen content, then assuming that all of this nitrogen belongs to protein, and multiplying the nitrogen value by a factor such as 5.7 or 6.3, which is the ratio of the mass of a polypeptide to the mass of the nitrogen atoms it contains. However, this procedure neglects the considerable proportion of seed nitrogen that belongs to the bases of the nucleic acids (Chapter 1). This imprecise method always provides a serious overestimate of protein content,[11,19,20,21] and yet it accounts for most of the published information. Someone relying on this method would not know whether an apparent increase in protein content for seeds of a transformed plant reflected a real increase in protein content or an increase in seed non-protein nitrogen.

Secondly, from reliable estimates of seed protein content and composition, it is evident that the total amount of protein in the seeds of a given species varies significantly with cultivar. The distribution of protein across the major solubility classes, namely albumins (water soluble proteins), globulins (salt soluble proteins), prolamins (alcohol soluble proteins) and glutelins (alkali or acid soluble proteins) also varies with species and with cultivar.[7] The proteins of legume seeds are mainly globulins and albumins, whereas the proteins of most cereals are mainly prolamins and glutelins. Thus it is already possible to choose varieties that have desirable protein constitutions in terms of the proportions found in these solubility classes, and consequently in their essential amino acid make-up.[22] There is little point in breeding for increased seed protein content *per se*, whether by conventional or molecular methods, unless some improvement in the representation of the first or second 'limiting' essential amino acid is achieved at the same time. These essential amino acids are identified in the next section.

ELEVATING ESSENTIAL AMINO ACID CONTENT IN SEEDS

Essential amino acids are those that cannot be synthesised by humans and other mammals, and hence they have to be ingested, mainly as protein. Most of the essential amino acids were mentioned in Chapter 3, because their synthesis is inhibited by various herbicides. Some have a branched carbon chain (valine, leucine, isoleucine); some have an aromatic ring (tyrosine, phenylalanine) or alternative complex ring structure (tryptophan); and others are polar (threonine) or have a net positive charge (lysine). Finally, the sulfur-containing amino acids methionine and cysteine are jointly considered essential, because the requirement for methionine can be reduced by the availability of cysteine.

Lysine, threonine and tryptophan tend to be lacking in the proteins of cereal grains, whereas the sulfur-containing amino acids are often lacking in legumes. There are striking exceptions, such as chickpea (*Cicer arietinum*), which has an excellent representation of the sulfur-amino acids and an adequate representation of all the other essential amino acids.[23,24] In chickpea and several other key legumes, it is the albumin fraction that is enriched both in sulfur-amino acid[23,25] and in tryptophan content.[24] Breeders seeking to manipulate the seed content of essential amino acids by gene transfer have so far concentrated on dicotyledons as subjects, rather than cereals, and have almost always attempted to elevate the methionine content by choosing to transfer the gene for a protein enriched in this amino acid.[26]

At CSIRO Plant Industry, narrow-leafed lupin (*Lupinus angustifolius*) has been modified by the addition of a gene for a sulfur-rich seed protein from sunflower.[27] The transformation was conducted via *Agrobacterium*. The methionine content of the seeds of the modified lupin is double that of the unmodified precursor variety — a 100 per cent increase. With the exception of white lupin (*Lupinus albus*), known also as 'lupini bean', lupin seeds are rarely eaten by humans because their content of bitter alkaloids makes them unpalatable. Seeds of other lupin species are fed mainly to animals, and this modified narrow-leafed lupin has been subjected to animal feeding trials in Western Australia, New South Wales, Denmark and Israel. It has proved particularly beneficial for poultry and sheep,[27] and should be ready for release in 2006. Originally an attempt was made to transfer the gene for this sunflower protein to clover, but the expression of the transferred gene in clover leaves was insufficient to make the development of this variety worthwhile.

REMOVING CAFFEINE FROM TEA AND COFFEE

In moderation, caffeine serves a useful function as a stimulant, allowing tired muscles and brains to work longer. In the mouth and upper digestive tract it also acts as an antibiotic, working generally against bacteria and fungi. People with high blood pressure (hypertension) are well advised to avoid excessive caffeine,[28] and a solvent-extraction procedure can be used to remove caffeine from coffee so that the flavour can be enjoyed without the adverse effects.

Professor Alan Crozier of Glasgow University has suggested that tea or coffee plants might be modified to produce less caffeine in the leaves or seeds, respectively. He has cloned a gene that controls one of the steps in caffeine synthesis, but has not yet achieved this modification. If successful, he would probably find the modified plants succumbing to some insect pest that might otherwise be resisted. This is what happened when the content of the phenolic compound gossypol was reduced in glandless cotton; the plants became more vulnerable to budworm (*Heliothis virescens*) and other insects.[29] Pest insects also gained the upper hand when bitter alkaloid content was reduced below 10 per cent of normal concentration in the stems and leaves of Western Australian lupin crops.[30,31]

USING FOODS TO CONVEY MEDICATION

There are many sound reasons why intramuscular injection is effective for conveying vaccine antigens to stimulate human immune systems. The Sabin oral vaccine for poliomyelitis is one exception. Now Charles J Arntzen and his colleagues want to produce edible vaccines in fruits, and a high priority has been suggested for hepatitis B vaccine in bananas.[32] The justification offered is that local production of the fruit might save the costs of refrigeration and transport presently incurred for conventional vaccines. This is a thoroughly impractical proposition. Trying to get any novel protein produced in sufficient concentration in a fruit would be an uphill battle, because fleshy fruits do not produce much protein except inside their seeds.[33] Bananas in any case have no seeds. A calculation similar to that done for the Antarctic fish serum glycoprotein proposal (Chapter 4) demonstrates the utter futility of the exercise.

For the foreseeable future, producing vaccines in fruits remains an impractical and inappropriate technology. A superior alternative is the vaccine patch, now being developed at the Centenary Institute of Cancer Medicine and Cell Biology in Sydney.[34]

USING PLANTS TO PRODUCE PLASTICS

Plastics have their place, although much more thought needs to be given to their re-use. There is really no need to engineer plants to produce plastics,[35] certainly not the kinds that resist microbial breakdown. No consideration has been given to what would happen to the accumulation of plant parts resistant to the normal pathways of decay. Conventional plant products are already being used to produce biodegradable plastics in the Netherlands, where the cost of burying wastes in landfill is considerable and a stimulus to the search for ecologically sensible solutions to problems of waste management. Plastics can also be substituted with more tractable materials. Maize starch was used to produce biodegradable plates, cutlery and bin liners during the Olympic Games in Sydney in September 2000. These were later collected and composted together with stable and other wastes.

REFERENCES
1   *The Australian*, 23 November 1988.
2   Beardsley, T (2000) Rules of the game. *Scientific American* 282(4): 24–25.
3   Conway, GR (2000) Crop biotechnology: Benefits, risks and ownership. Paper delivered at OECD Conference on *GM Food Safety: Facts, Uncertainties and Assessment*. 28 March, Edinburgh.
4   Ye, XD, Al-Babili, S, Kl'ti, A, Zhang, J, Lucca, P, Beyer, P & Potrykus, I (2000) Engineering the complete provitamin A (beta-carotene) biosynthetic pathway into (carotenoid-free) rice endosperm. *Science* 287: 303–05.
5   ABC Radio News, 13 September 2000.
6   Scrimshaw, NS & Young, VR (1976) The requirements of human nutrition. *Scientific American* 235(3): 50–64.
7   Murray, DR (1984) Accumulation of seed reserves of nitrogen. In DR Murray (ed.) *Seed Physiology Volume 1. Development*. Academic Press, Sydney, pp. 83–137.
8   Murray, DR (1997) Effects of [CO_2] on seed yield and quality. In DR Murray *Carbon Dioxide and Plant Responses*. Research Studies Press, Taunton, UK, pp. 177–94.
9   Azhakanandam, K, Power, JB, Lowe, KC, Cocking, EC, Tongdang, T, Jumel, K, Bligh, HFJ, Harding, SE & Davey, MR (2000) Qualitative assessment of

aromatic indica rice (*Oryza sativa* L.): Proteins, lipids and starch in grain from somatic embryo- and seed-derived plants. *Journal of Plant Physiology* 156: 783–89.
10 Bingley, W (1831) *Useful Knowledge — Volume II. Vegetables*, 5th edn. Baldwin & Cradock, London.
11 Murray, DR (1999) Eating peas and beans. In DR Murray *Growing Peas and Beans*. Kangaroo Press, Sydney, pp. 59–65.
12 Lott, JNA (1984) Accumulation of seed reserves of phosphorus and other minerals. In DR Murray (ed.) *Seed Physiology Volume 1. Development*. Academic Press, Sydney, pp. 139–66.
13 Murray, DR (1990) Deleterious consequences of eating irradiated foods. In DR Murray *Biology of Food Irradiation*. Research Studies Press, Taunton, UK, pp. 137–82.
14 Eggum, BO (1981) Nutritional problems related to double low rapeseed in animal nutrition. In ES Bunting (ed.) *Production and Utilization of Protein in Oilseed Crops*. Martinus Nijhoff Publishers, The Hague, Boston, London, pp. 293–310.
15 Buzza, G (1991) Canola. In RS Jessop and RL Wright (eds) *New Crops — Agronomy and Potential of Alternative Crop Species*. Inkata Press, Melbourne and Sydney, pp. 19–26.
16 Stryer, L (1981) *Biochemistry*, 2nd edn. WH Freeman & Company, San Francisco.
17 Slack, CR & Browse, JA (1984) Synthesis of storage lipids in developing seeds. In DR Murray (ed.) *Seed Physiology Volume 1. Development*. Academic Press, Sydney, pp. 209–44.
18 Green, A (2001) personal communication, 3 May.
19 Murray, DR (1986) *Guidelines for the Quantitative Estimation of Protein in Plant Tissue Extracts*. Department of Biology Technical Report No. 4, The University of Wollongong, Wollongong, NSW.
20 Murray, DR & McGee, CM (1986) Seed protein content of Australian species of *Acacia*. *Proceedings of the Linnean Society of NSW* 108(3): 187–90.
21 Murray, DR (1997) Ecological implications of the bush tucker restaurant industry. In *Sustainable Use of Wildlife: Utopian Dream or Unrealistic Nightmare?* Nature Conservation Council of NSW, Sydney, pp. 146–69.
22 Murray, DR (1990) Production of microbial toxins in stored irradiated foods and loss of quality in wheat. In DR Murray *Biology of Food Irradiation*. Research Studies Press, Taunton, UK, pp. 113–35.
23 Murray, DR & Roxburgh, CMcC (1984) Amino acid composition of the seed albumins from chickpea. *Journal of the Science of Food and Agriculture* 35: 893–96.
24 Sastry, MCS & Murray, DR (1986) The tryptophan content of extractable seed proteins from cultivated legumes, sunflower and *Acacia*. *Journal of the Science of Food and Agriculture* 37: 535–38.
25 Murray, DR (1979) A storage role for albumins in pea cotyledons. *Plant, Cell and Environment* 2: 221–26.
26 Xue, G-P, Perroux, JM, Patel, M & Johnson, JS (1999) Prospects for improvement of the nutritive value of cereal protein by molecular breeding. *Proceedings 11th Australian Plant Breeding Conference*, CRC for Molecular Plant Breeding, Adelaide, Volume 2: 176–77.
27 Higgins, TJV (2001) personal communication, 3 May.
28 Boyce, Nell (2000) Storm in a coffee cup. *New Scientist* 165(No. 2223): 28–31.
29 Gatehouse, JA (1991) Breeding for resistance to insects. In DR Murray (ed.) *Advanced Methods in Plant Breeding and Biotechnology*. CAB International, Oxford, pp. 250–76.

30 Bell, EA (1984) Toxic compounds in seeds. In DR Murray (ed.) *Seed Physiology Volume 1. Development*. Academic Press, Sydney, pp. 245–64.
31 Davidson, BR & Davidson, HF (1993) Legume crops. In BR Davidson and HF Davidson *Legumes — The Australian Experience*. Research Studies Press, Taunton, UK, pp. 346–79.
32 Langridge, WHR (2000) Edible vaccines. *Scientific American* 283(3): 66–71.
33 Murray, DR (1988) Fruit development. In DR Murray *Nutrition of the Angiosperm Embryo*. Research Studies Press, Taunton, UK, pp. 95–119.
34 Thornton, M (2001) No more doctors' needles. *Sun-Herald*, 18 February, p. 41.
35 Shewry, PR, Napier, JA & Davis, P (eds) (1998) *Engineering Crop Plants for Industrial End Uses*. Portland Press, London.

6
ENVIRONMENTAL AND HEALTH IMPACTS OF GENETICALLY MODIFIED PLANTS

> For fools rush in where angels fear to tread
> Alexander Pope

WHY WORRY?

Large multinational companies that have a financial interest in the development of genetically modified plants often make contradictory statements about what they are doing. In one breath, they are achieving marvellous things for humanity with revolutionary techniques. But in the next, when they want to conceal what they have done by refusing to label the products, they claim they have not done anything very different after all. We are told that plant improvement has been going on for thousands of years, and that these newly developed transgenic specimens are just the latest in a long line of innovations.

This pretext is transparent. The new plants *are* significantly different from the products of previous breeding because of the imprecision of gene placement and expression inherent in the latest transformation procedures, and because of reliance on bacterial and viral DNA sequences that accompany the transferred genes (Chapter 2). Bacterial proteins are being produced in the edible parts of genetically modified plants, because bacteria have been the sources of reporter genes as well as functional genes, such as Bt-proteins and barnase (Chapter 4). Consuming significant quantities of proteins from bacteria not normally associated with the digestive tract is not something humans should embark upon without due care. For this reason alone we need to be extremely cautious about transgenic plants and their products. Risk assessment has become an important task facing regulatory bodies throughout the world.

When health issues have been raised about genetically modified plants in the past, they have often been dismissed. At the time of the first Australian Parliamentary Inquiry,[1] the Showa-Denko tryptophan affair (see Chapter 2) was played down and the claims of ill-health or death on the part of the victims were said to be 'disputed'. With regard to Bt-proteins, the potential for human toxicity was ignored. There is now no doubt that these bacterial proteins can cause degenerative changes in the cells lining the small intestine (ileum) in mammals.[2] Few would agree that genetically modified plants pose 'no perceptible risk to public health', the sublimely optimistic position taken by Professor Adrienne Clarke.[3]

Earlier false assurances underline the need for rigorous testing now and in the future. However, we should not run to the opposite extreme and ban every genetically engineered plant solely because of the way it was produced. The precautionary principle needs to be applied sensibly. Every proposed release of a novel genetically modified plant should be judged on its merits after suitable field trials, feeding tests and public consultation. The risks to the environment or to human or animal health will vary according to the nature of the modification and the way it was achieved. How well these risks are assessed in advance will determine the ultimate general acceptance or rejection of gene technology for plants.

ENVIRONMENTAL RISK ASSESSMENT

Concerns about the escape of genetically modified organisms expressed during the mid-1970s led to the formulation of guidelines for physical and biological methods of containment for experiments involving recombinant nucleic acid techniques.[4] However, there have since been some startling releases of genetically modified plants without proper appreciation of the obvious consequences. This has occurred mainly in the United States and Canada, which together account for more than 80 per cent of releases.[5] Genetically modified plants should not have been treated as 'substantially equivalent' to unmodified varieties, as they have been in these countries, because they are not.

Some people advocate modelling to forecast consequences,[6] but modelling cannot help if the critical risks are not perceived in the first place. Models also need to be tested by experiment. Who is going to pay for independent research of this kind? In the absence of empirical information, risk assessment should remain conservative and err on the side of caution.

There should always be an assessment of the likelihood of genetic escape from any released transgenic plant, and an assessment of the effects the introduced genes might have if transferred beyond the variety that has been released, or indeed to other kinds of organism, such as the bacteria and fungi that are normally involved in the breakdown of dead plant material in the soil. Genetic escape should remain the responsibility of whichever company claims ownership of the variety from which escape occurred, and damages should be paid to growers whose cultivars are altered by undesirable genetic exchange with released transgenic plants.

Taking a broader perspective, environmental damage will be caused by the kind of growing methods encouraged by the adoption of genetically modified plants. Vast and repetitive monocultures are the norm for maize, soybean and cotton. Spacing these crops to suit mechanical harvesters allows unsustainable rates of soil erosion to occur, for example, the rates reported in Canada.[7] A loss of one inch of topsoil translates to a decrease in maize yields of 30 to 40 per cent, or for wheat, a reduction of 1.5 to 3.4 bushels per acre.[7] Net losses of 20 tons of soil per acre (8.2 tonnes per hectare) every year are commonplace in Canada.[8] Such losses deplete the soil of its content of organic matter with associated beneficial bacteria and fungi, and take away nutrients that then have to be replaced with expensive fertilisers. This degree of wastage is in addition to the routine overuse of synthetic nitrogenous fertilisers, which leads to significant contributions to greenhouse gas emission and eutrophication of waterways from solutes that cannot be used by the crop or held within the soil. The current practice in Iowa typifies this problem. The soil is gassed with ammonia in autumn, seven months before sowing, and consequently most of the applied nitrogen is lost before sowing occurs.[9]

Large populations of identical plants encourage pests and diseases, and the use of pesticides. As with herbicides (Chapter 3), most of what is applied is wasted, either spread too widely during application or washed away soon afterwards. Pesticides are rarely highly specific, killing countless other beneficial organisms on the plants themselves, or in the soil and groundwater.

Land clearance for large scale agriculture reduces the capacity of native vegetation to fix carbon dioxide in a cumulative way. This is a vicious circle. Allowing atmospheric carbon dioxide concentration to continue to increase will lead to the erosion of nutritional quality in the very foods being produced.[10] The displacement of better ways of growing plants in many parts of the world is encouraged by the same companies that promote genetically modified crop plants (see Chapter 8).

THREATS TO HUMAN OR ANIMAL HEALTH

As indicated in Chapter 4, potentially insecticidal proteins and glycoproteins in legume seeds have been consumed safely for thousands of years, provided the seeds are cooked. Transgenic plants producing these same proteins and glycoproteins in edible parts other than the seeds should therefore be safe to eat after cooking, but complications might follow when adverse effects are produced by genes accompanying those intended for transfer (see below). Furthermore, a significant proportion of any population may be sensitive to the protein that is introduced, such as the Brazil nut protein transferred to soybean[11,12] and the Bt-protein transferred to StarLink maize (see below).

THE NEED FOR 'EMPTY VECTOR' EXPERIMENTS

The potato variety Desiree was modified to produce the lectin from snowdrop (*Galanthus nivalis*) as a defensive measure against aphids and nematode worms (Chapter 4). This potato was tested in animal feeding experiments because the snowdrop lectin is not part of any customary human diet. Rats were fed with transformed potato, unmodified potato to which purified lectin had been added, and unmodified potato, and this was done with both raw and cooked potato.[13] Assay showed that cooking for one hour reduced the active lectin content by at least 80 per cent. After 10 days of the potato diet, the rats were killed and histological samples of stomach, jejunum, ileum, caecum and colon were prepared. The thickness of the gastric mucosa was measured for the stomach samples, wall thickness was measured for the caecum, and 'crypt length' was measured for the other samples. This measure reflects the presence of invaginations, or villi, throughout the small intestine (jejunum plus ileum) and the large (the colon).

The stomach wall was significantly thicker when rats were fed potato diets including the snowdrop lectin (Table 6.1). Transformed potatoes and unmodified potatoes to which lectin had been added gave similar results and cooking made no difference. In the jejunum, the crypt length for rats fed raw transgenic potato was significantly greater than for rats fed unmodified potato or unmodified potato to which lectin had been added. Boiling the potatoes abolished this effect.

For the ileum, there were no effects that could be attributed either to the lectin or to transformation. Rats fed boiled potato of any kind (unmodified, lectin-added, or transformed) had significantly shorter ileal crypts than rats fed uncooked potato of any kind. With regard to the caecum, rats fed boiled transgenic potato had significantly thinner mucosa than rats fed either of the control diets boiled or any of the three diets uncooked.

Table 6.1
Mean wall thickness (stomach, caecum) or crypt length (jejunum, ileum, colon) for rats fed transgenic potato or control diets,[13] expressed as micrometres (SD)

		Unmodified Desiree		Unmodified Desiree + lectin gene		Transgenic potato expressing lectin	
Stomach	boiled	294	(46)	347	(42)	339	(36)
	raw	261	(32)	312	(32)	323	(54)
Jejunum	boiled	75	(19)	78	(17)	78	(12)
	raw	57	(8)	64	(11)	90	(20)
Ileum	boiled	59	(8)	55	(7)	63	(13)
	raw	71	(9)	79	(13)	87	(25)
Caecum	boiled	95	(19)	98	(21)	70	(15)
	raw	132	(19)	104	(17)	119	(25)
Colon	boiled	146	(15)	177	(24)	139	(24)
	raw	192	(34)	148	(25)	215	(34)

The enhanced growth in the jejunum resulting from the diet of raw transgenic potato is of great interest. The obvious conclusion, drawn correctly by these two researchers,[13] is that the lectin itself is not responsible for all the observed effects of eating the transgenic potato. Alternative explanations must be sought, and it is reasonable to suggest that some component of the construct other than the lectin gene might be responsible.

Various uninformed comments have been circulated about this study, some of them emanating from Sir Robert May under the auspices of the Royal Society.[14] Although Sir Robert May is an eminent scientist, he is a physicist and has no qualification in biological sciences. Yet he takes it upon himself to criticise these researchers unduly. Such appeals to eminence and authority rather than reason are quite unnecessary. Science is an empirical enterprise. We need only ask whether the findings of these researchers can be confirmed by others acting independently.

One criticism of this study being transmitted verbally is that the potatoes were green because of undue exposure to light. This would mean that they contained the alkaloid solanine. Given that the researchers should have been aware of basic precautions to take when storing potatoes, and that Desiree has a pigmented skin, this explanation lacks plausibility. A comment that does merit consideration in further research comes from Harry Kuiper and colleagues,[15] concerning the need for preparation of a transformed plant with an 'empty vector'.

In other words, the same variety of potato should be transformed, but the gene for the snowdrop lectin should be excluded from the full construct. This is a good idea, since the effects of consuming transformed potatoes on rat digestive tract parameters that could not be attributed to the lectin *per se* should still be observable in response to the transformation. Those making this suggestion have experience in its application to the assessment of a transgenic tomato expressing the gene for one of the Bt-proteins, CryIA(b).[16] It should be further noted that a plant transformed with an empty vector would very probably have the transgene package in a different position (or positions) compared with the plant already transformed with the full construct (Chapter 2), and this difference would need to be considered when interpreting results.

MISUSING BT-PROTEINS

Widespread publicity has been given to the observation that pollen from maize plants genetically modified to produce a Bt toxin can kill the larvae of monarch or wanderer butterflies (*Danaus plexippus*). This can happen when the pollen is dusted onto the surface of their usual food plant, milkweed (*Euphorbia heterophylla*), as shown under laboratory conditions by John Losey and Linda Raynor.[17] Maize pollen is normally dispersed by wind from male flowers held at the tops of the plants. So much maize is grown that vast clouds of pollen can rise as male flowers mature, ensuring a wide distribution of pollen in those maize-growing areas of the United States that are home to the monarch butterfly. If these results extrapolate from the laboratory to the real world as expected, then adverse ecological consequences would extend to organisms in a food chain that depend on the consumption of healthy monarch caterpillars (some birds) or adults (bats),[17] as fewer larvae will mature.

There is no difference in principle between allowing transgenic maize plants to release insecticidal pollen and spraying an insecticide. The undesirable ecological results are the same. It is appalling that a transformed maize plant has been released that does not express its Bt gene specifically — in leaves and stems, for example, rather than in every part of the plant. This lack of specificity in expression is a glaring fault and should have been sufficient reason to prohibit the release of this variety.

Another reason that Bt-transformed plants should not be released so readily in future is that individual Bt-proteins can cause allergenic reactions in humans. For this very reason, StarLink corn (developed by Aventis) was approved in the United States for production of animal food. StarLink was not approved for human consumption. But a mix-up occurred. StarLink maize was detected in a battery of products from several companies, including cornflakes from Kellogg and taco shells from Taco Bell.[18] Recalls extended to batches of contaminated maize products that had already entered several other countries, and cost millions of dollars.

Bt-transformed potatoes have not passed their feeding trials with flying colours either. A Draft Risk Analysis report prepared by the Australia New Zealand Food Authority (ANZFA)[19] refers to a one-month feeding trial in which rats were each given one fresh raw potato every two to three days of the study. The transformed variety, New Leaf Russet Burbank (line BT-06), was compared with Russet Burbank controls. 'Rodent Chow' was continuously available to both groups of rats, so that potato was not the sole source of nutrition. Although cumulative weight gains were normal, 'gross pathology revealed a number of abnormal findings, such as enlarged lymph nodes, hydronephrosis and enlarged adrenals'. Hydronephrosis is a disorder of the kidney. This assessment[19] goes on to dismiss these findings as 'observed in both the test and control groups', without stating unequivocally whether the incidences were unequal or equal in both groups. This is most unsatisfactory, as the exact amounts of potato ingested by each animal were not measured, nor was the effect of cooking the potato tested. The experiment should have been repeated more stringently, but the obvious need for further studies was ignored.

With a little more basic research, the genes for Bt-proteins could probably be modified so that they produce proteins that are not allergenic to humans or animals, while retaining their effectiveness against pest insects. There is no technical barrier to hinder this approach.

Irrespective of food safety issues, a third reason to prohibit the release of plants transformed to express only a single Bt-protein is that target organisms will be encouraged to become resistant to whichever Bt-protein is being produced. This is the same selection principle that operates whenever a new synthetic pesticide or herbicide is used. The next generation comes from survivors of the first exposure, and these are more likely to carry heritable resistance. Resistance is then amplified in just a few subsequent generations.

It has often been suggested that pest organisms are more likely to develop resistance to Bt-proteins because the production of these proteins in leaves or other plant parts provides continuous rather than intermittent exposure. This is not the reason. Resistance is far more likely to be engendered because a complex mixture of toxic proteins that interact synergistically (Chapter 4) has been greatly simplified by gene technologists. The genes for Cry proteins are being transferred and offered to pest species one at a time, for example StarLink maize received the gene to produce protein CryIXC. This overly simplistic approach is asking for trouble. When resistance to one Bt-protein has been elicited, the technologists will have to move on to the next, and so on. Eventually 'superpests' could result, each with a master set of resistances to a full spectrum of Bt-proteins.

The same outcome would not result from continued use of whole *Bacillus thuringiensis* preparations such as Di Pel. This form of bacterial insecticide has been in use for at least 70 years without any such problems, because it is highly improbable that a target organism could develop simultaneous resistance to all the components of a mixture of these proteins. However, the future use of preparations like Di Pel will be undermined by gene technology unless regulators become more alert to basic biological principles.

Fortunately, steps are being taken to increase the number of Bt-protein genes expressed simultaneously. New cotton plants expressing two different *cry* genes are currently being developed by CSIRO Plant Industry.[20] The protein combinations will be CryIA(c) and CryIIA, or CryIA(c) and CryX. This is a move in the right direction. The cotton plants producing a single Bt-protein that were released in Australia in 1996 (Monsanto's INGARD and CSIRO varieties produced under licence by crossing) are not as effective against Australian insects as INGARD is against pest insects in the United States. Although the use of pesticides on Bt-cotton in Australia has fallen by an estimated 50 per cent relative to non-Bt cotton, this reduction is confined to the first half of the growing season. The expression of the *cry* gene declines in the second half of the growing season, when growers revert to pesticide use just as if they were growing unmodified varieties. Australian growers of Bt-cotton still use 12 to 14 applications of pesticides, whereas American growers use three or four applications.[20]

HOW SAFE IS TERMINATOR?

The various components of Terminator as applied to seeds (Chapter 4) would be transmitted by natural crossing of genetically modified varieties with unmodified varieties, leading to a general deterioration in seedling survival from affected seeds. Given the importance of high germination rate and seedling vigour to agriculture in general, it does not seem wise to encourage the possibility of this kind of haphazard impairment.

Whether it is safe to consume barnase present in harvested seed crops is a separate issue. We cannot assume that it is safe for everyone to consume bacterial proteins, especially those with powerful toxic or catalytic effects, such as barnase. A great deal more information is needed from animal and human feeding trials. The participation of human subjects should be voluntary, before release, rather than obligatory, after release.

HOW SAFE ARE VIRAL NUCLEIC ACID SEQUENCES?

Concern is mounting about the use of viral nucleic acid sequences as promoters in the constructs that enable a transgene to be introduced to plant cells. Viruses that subsequently infect plants derived from such cells may preferentially combine with these promoter regions, with possibly dire consequences. Professor Adrian Gibbs, a virologist formerly from the Australian National University, Canberra, has often warned of the possibility of plant viruses becoming altered to the point where they might infect unrelated plants, animals or humans. It is in the nature of viruses to mutate and constantly shift their virulence or their host. Furthermore, sudden and radical host switches have been documented in viral evolution. We should not be surprised if something like this happens with some of the genetically modified plants that have already been released.

The most commonly used viral promoter is the CaMV35S sequence, from the cauliflower mosaic virus. A major drawback has already been noted — the lack of tissue or organ specificity for the expression of genes associated with this promoter. Moreover, the CaMV promoter sequence has been identified as already present within plant DNA.[21] Surely this supports concern about the interactive nature of such sequences, and underlines the potential for viral sequences to be adopted by viruses infecting cells that possess these sequences? Further confirmation that this is possible comes from evidence that cauliflower mosaic virus itself can inactivate transgenes under the control of the CaMV35S promoter.[22]

The use of other viral sequences can be ranked from those with greatest risk (satellite sequences) to those with least risk (non-transcribed regions of a viral genome; Chapter 4). The use of viral sequences that result in destruction of invading viral particles is the best option. If a virus does not reproduce following invasion of appropriately modified plant cells, then that particular virus will be unable to acquire or transmit any new sequences. We need to be sure this is always going to happen.

GENES FOR ANTIBIOTIC RESISTANCE

Excessive use of antibiotics as prescribed medications and in animal feed supplements has encouraged an increase in resistance among many disease-causing bacteria. This resistance is both inherited and transferable, as it resides outside the central bacterial chromosome in the plasmids. Through the process of conjugation, bacteria are constantly exchanging plasmids and the genetic information contained in them.[23] Multiply-resistant wound or food-poisoning bacteria such as golden staph (*Staphylococcus aureus*) are now very common, especially in hospitals.[24]

In practice, this means that specific antibiotics are less effective than when they were introduced, and cures relying on the use of antibiotics are far more uncertain. Conditions like stomach ulcers, caused by *Helicobacter pylori*, are less easily treated,[25] as populations of this bacterium now display substantial resistance to two of the three antibiotics used simultaneously to cure gastric ulcers since 1986.

To preserve the use of antibiotics for medical emergencies, the opportunities for transfer of antibiotic resistance genes among bacteria should be minimised. Antibiotic resistance genes used as reporter genes in transgenic food plants might be transferred to other organisms, especially bacteria, during digestion. This is sometimes referred to as 'horizontal' gene transfer, and it has been shown to happen. Bees are able to mediate transfer of genes from the pollen of genetically modified canola to the bacteria that reside in their gut, according to research conducted by Professor Hans-Heinrich Kaatz in Jena. Moreover, the detached leaves of four diverse genetically modified plants were able to pass their genes for antibiotic resistance to the fungus *Aspergillus niger*[26,27] cultured on the leaves.

It would be preferable if antibiotic resistance genes were no longer used as reporter genes in the process of propagating transgenic plants. It is possible to substitute markers that do not involve antibiotic resistance. Some confer the ability to use unusual nutrients, such as the sugar mannose. This ability is the basis of Syngenta's Positech system, patented in March 2000.

HOW SAFE ARE TRANSPOSABLE ELEMENTS?

Some DNA sequences have the propensity to move within a chromosome and even from chromosome to chromosome. These so-called 'jumping genes' were first identified in the banana fly *Drosophila melanogaster*, and then shown to occur in maize by Barbara McClintock.[28] One such sequence, known as 'mariner', has been identified in diverse species and is attractive to gene technologists as a means of transforming plants.[29]

Given the nature of such sequences, and the fact that they have already moved from one species to another fairly often, the likelihood of unintended gene transfers is very high. Dan Hartl, who discovered 'mariner' in the 1980s, recommends that we tread carefully[29] with regard to its use for genetic transformation.

A FILTER QUESTIONNAIRE

Regulatory authorities could fulfil their function more capably if they applied a simple screening questionnaire, such as the following:

1. Does the genetically modified plant express a gene from another plant species, or another organism, so that a potentially allergenic protein or glycoprotein is produced in the edible parts of the plant?
2. Is the genetically modified plant going to be subjected to a systemic herbicide that will be carried into the edible parts of the plant?
3. Is the genetically modified plant capable of transmitting any other kind of systemic pesticide into edible parts of the plant?

4   Do any of the concomitant transgenes confer resistance to an antibiotic still useful in the treatment of human or animal diseases?
5   Are any of the transferred genes expressed in the pollen grains?
6   Does the promoter region include a viral DNA sequence that could be captured by viruses infecting the plant?

A single 'yes' answer should bring immediate rejection; 'not known' should bring a request for more research to find out. If all questions are answered 'no', then the plant could be considered further. Localisation of the transferred genes could be the next step. Very few of the genetically modified plants released to date would pass this questionnaire.

This is not like a medical situation, where failure to intervene could mean death and a calculated risk may reasonably be taken. There should be no hurry to replace reliable varieties of food plants with suspect alternatives. Apart from the obvious, 'Particular attention must be given to the detection and characterisation of unintended effects of genetic modification.'[15] A variety of canola modified to be resistant to the herbicide glyphosate (GT73) provides an illustration of an unintended effect of modification. Rats fed on meal of this transgenic canola compared with an unmodified variety at 15 per cent of total dietary intake displayed higher liver weights (12 to 16 per cent), an adverse effect that was attributed to higher contents of glucosinolates.[30] The genetically modified variety had 4 grams per kilogram glucosinolate content, whereas the unmodified variety had 1.8 grams per kilogram. Because canola meal is not regarded as fit for human consumption, this disparity was overlooked by ANZFA, who stated that 'The higher level of glucosinolates present in glyphosate-tolerant canola was not attributed to the genetic modification.' This is illogical, and shows that our committees really do need to take their responsibilities for food safety more seriously. Clearly these adverse effects on rats are consistent with the effects noted long ago for other animals and poultry,[31] underlining the fact that canola meal is not fit for animal consumption either (see Chapter 5).

One way of discovering the unintended effects of genetic modification is by profiling gene expression, organ by organ, or over time. This is possible with a new probe system, termed 'microarray', which uses immobilised DNA and can examine the expression of hundreds of genes at once.[15,32] Extra testing would add to the expense of developing new genetically modified varieties, but this cost would no doubt be recovered from consumers.

THE 'ALIAS' STRATEGY

There are alternative ways of attaining the same breeding goals without relying on problematic transformation procedures or engaging in animal feeding experiments that can also be objected to on the grounds of cruelty. Keeping to the standard procedures of crossing and selection, while using some of the techniques of nucleic acid technology to gain information, is the approach adopted successfully by the University of Sydney Plant Breeding Institute (Chapter 7) and by some members of CSIRO Plant Industry and their collaborators in the area of wheat improvement.

It is possible to use the techniques of recombinant DNA technology in a way that does not result in the release of any transgenic plants. Selected plants can be modified to find out what role individual genes have in the plant's metabolism, physiology, defence systems or gross composition. Changes in gene expression during a developmental time-course can be monitored by microarray probe techniques,[32] confirming or negating proposed gene functions.

Varieties from existing collections can then be screened to identify those that most closely resemble the experimental transgenic variety. When the transgenic plants are no longer needed, they can be collected and destroyed. This is a normal precaution when growing transgenic plants in glasshouses. The plants that are released, or used for further conventional breeding, are not transgenic; they are plants that existed before the transgenic experiments were undertaken.

Many varieties of wheat have been bred in Australia by conventional procedures.[33] Yet there are still improvements to be made. The qualities of wheat grain that relate to baking properties of milled flour include the distribution of starch granules between the two categories A and B. Type A granules are larger, initiated early during seed development. Type B granules are smaller, initiated later during seed development, and they are easily rinsed out during processing of milled grain, causing problems of disposal. A desirable genotype would have more type A granules and fewer type B, although some Type B granules are needed as they contribute to the extensibility of the dough.

Another breeding goal for wheat is to find or produce varieties with low amylose content. Amylose is starch that consists largely of a linear polymer of glucose, forming a helix. The alternative form of starch is amylopectin, which is amylose with branch-points, a more complex molecule. Starch consisting mainly of amylopectin is also termed 'waxy' and has properties that are desirable for some products.

The 'alias' strategy is being applied to both of these goals by Dr Matthew Morell from CSIRO Plant Industry and his colleagues.[34] A two-fold advantage is gained by this strategy. First, because the varieties to be released have been selected from plants that had already been bred by conventional means, there are no extra expenses or impediments from intellectual property wrangles (Chapter 7). Secondly, any potential risks associated with the technology have been avoided, and the products of such plants will not be rejected by consumers because of the way in which the plants were developed.

1
White flowers can result from simple recessive mutations that prevent pigment synthesis (pea, Stella Nova).

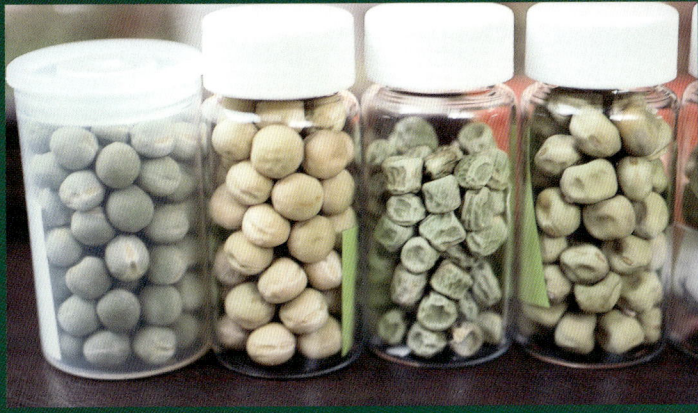

2
Mendelian seed characters are readily distinguished: round vs wrinkled seed shape, and yellow vs green embryo.

3 *below*
Snowdrop (*Galanthus nivalis*), controversial source of the gene for a defensive lectin.

4
Desiree potato, a subject for transformation to produce snowdrop lectin, has useful skin pigmentation that prevents greening from incidental light.

5 *above*
Canola — the crop that is also a weed.

6
CSIRO Plant Industry's transgenic lupin shows a doubling of seed methionine content.

7 *above*
Transgenic cotton plants at full height inside a PC2 glasshouse at CSIRO Plant Industry, Canberra.

8 *above*
Guidelines for containment and disposal are prominently displayed in CSIRO facilities.

9
Rye plants are a source of useful characters for conventional wheat breeding at the University of Sydney's Plant Breeding Centre.

10
Anigozanthos Bush Sunset, produced by conventional breeding, on display at the Australian National Botanic Gardens, Canberra.

11 Banksia Birthday Candles, a dwarf form for which Plant Breeder's Rights were awarded.

12 *inset* Cultivar names should not be given to plants like this waratah unless they are distinct from plants still growing naturally.

13 Marguerite daisies undergoing crossing at the University of Sydney's Plant Breeding

14 *Scaevola* New Wonder at the University of Sydney's Plant Breeding Centre.

15 *above*
Two new petunias at the University of Sydney's Plant Breeding Centre.

16 *inset left*
Paper daisies undergoing field trials at the University of Sydney's Plant Breeding Centre.

17 *below*
Bracteantha (paper daisy) with unusual colour at the University of Sydney's Plant Breeding Centre.

18 *inset left*
The flower of the lotus is eaten as a vegetable in Malaysia.

19 *above* Climbing beans with 'tepee' supports at the Titi Organic Farm, Malaysia.

20 Cobs of the Vietnamese cold-tolerant maize variety Sapa laid out to dry; note the lack of uniformity in colour.

21 A 'Save the Seeds Campaign' conference in Nepal attended by the Chipko group, already famous for having saved their trees.

22 This woman is the seed guardian in the village of Gumi, Western Nepal. She supplies seeds to anyone who needs them.

23
The local radish cultivar in Nepal is highly favoured for pickling.

24 *right*
Packets of seeds are attached to fruit and vegetables for sale in Nepal.

25 *below*
In Maharashtra, India, staff of the Academy of Development Sciences curate a collection of more than 400 varieties of paddy rice, and grow them all out every year.

26
Varieties of millet, paddy rice and grain legumes belonging to a village seed bank in Tamil Nadu, South India.

27
A workshop on seed saving in Tonga, conducted by Michel Fanton.

28 *right*
To promote preservation of local plant varieties in the Solomon Islands, the Planting Materials Network has a stall at the local market.

29
In the Solomon Islands, seed viability cannot be prolonged for more than a few months.

REFERENCES

1. *Genetic Manipulation: The Threat or the Glory?* Report by the House of Representatives Standing Committee on Industry, Science and Technology, Canberra.
2. Fares, NH & El-Sayed, AK (1998) Fine structural changes in the ileum of mice fed on δ endotoxin-treated potatoes and transgenic potatoes. *Natural Toxins* 6: 219–33.
3. SBS Television (2000) 'Insight', 8 June.
4. Ada, GL (1979) Genetic engineering research: Evaluation and containment of potential risks. In FWG White (ed.) *Scientific Advances and Community Risk*. Australian Academy of Science, Canberra, pp. 7–20.
5. Cooper, W & MacLeod, J (1999) Genetically modified crops: the current status. In S Andrews, AC Leslie and C Alexander (eds) *Taxonomy of Cultivated Plants: Third International Symposium*. Royal Botanic Gardens, Kew, UK, pp. 277–83.
6. Burgman, MA (1999) Are Australian standards for risk analysis good enough? *Australian Biologist* 12: 125–37.
7. Standing Senate Committee on Agriculture, Fisheries and Forestry (1984) *Soil at Risk — Canada's Eroding Future*. The Senate of Canada, Ottawa, Canada.
8. Hill, S (2001) personal communication, 23 October.
9. Murphy, C (2001) Travel report — WSSA meeting and mid-west USA visit, February 2001. *A Good Weed* 23: 6–8.
10. Murray, DR (1997) Effects of [CO_2] on seed yield and quality. In David R Murray *Carbon Dioxide and Plant Responses*. Research Studies Press, Taunton, UK, pp. 177–94.
11. Conway, GR (2000) Crop biotechnology: Benefits, risks and ownership. Paper delivered at OECD Conference on *GM Food Safety: Facts, Uncertainties and Assessment*, 28 March, Edinburgh.
12. Specter, M (2000) The food that bit back. *Good Weekend*, the *Sydney Morning Herald* Magazine, 10 June, pp. 18–24.
13. Ewan, SWB & Pusztai, A (1999) Effects of diets containing genetically modified potatoes expressing *Galanthus nivalis* lectin on rat small intestine. *The Lancet* 354: 1353–54.
14. Horton, R (1999) Genetically modified foods: "Absurd" concern or welcome dialogue? *The Lancet*, 354: 1314–15.
15. Kuiper, HA, Noteborn, HPJM & Peijenburg, ACM (1999) Adequacy of methods for testing the safety of genetically modified foods. *The Lancet* 354: 1315–16.
16. Noteborn, HPJM, Beinenmann-Ploum, ME & van den Berg, JHJ (1999) Safety assessment of the *Bacillus thuringiensis* insecticidal crystal protein CryIA(b) expressed in transgenic tomatoes. In K-H Engel, GR Takeoka and R Teranishi (eds) *Genetically Modified Foods: Safety Issues*. ACS Symposium Series 65, Washington DC, pp. 134–47.
17. Baskin, Y (1999) Into the wild. *Natural History* 10/99: 34–37.
18. Anonymous (2000) Modified corn alert forces Kellogg to close plant. *Sydney Morning Herald*, 23 October, World 11 (from the *Washington Post*).
19. ANZFA Draft Risk Analysis Report. Application A382. Food derived from insect-protected potato lines BT-06, ATBT04-06, ATBT-0431, ATBT04-36, and SPBT02-05. Australia New Zealand Food Authority, Canberra.
20. Llewellyn, D (2001) personal communication, 3 May.
21. Higgins, TJV (2000) personal communication, 17 July.
22. Al-Kaff, NS, Kreicke, MM, Covey, SN, Pitcher, R, Page, AM & Dale, PJ (2000) Plants rendered herbicide-susceptible by cauliflower mosaic virus-elicited suppression of a 35S-promoter-regulated transgene. *Nature Biotechnology* 18: 995–99.

23 Stryer, L (1981) *Biochemistry*, 2nd edn. WH Freeman & Company, San Francisco.
24 Goodyer, P (2000) Bugs bite back. *Sunday Life!* The *Sun-Herald* Magazine, 17 September, p. 31.
25 ABC Television (2002) 'Health Dimensions'. 5 March.
26 Hoffmann, T, Golz, C & Schieder, O (1994) Foreign DNA sequences are received by a wild-type strain of *Aspergillus niger* after co-culture with transgenic higher plants. *Current Genetics* 27: 70–76.
27 Anderson, L (2000) What is genetic engineering? In L Anderson *Genetic Engineering, Food and our Environment*. Scribe Publications, Melbourne, pp. 19–31.
28 Peters, JA (ed.) (1959) *Classic Papers in Genetics*. Prentice-Hall, Englewood Cliffs, New Jersey.
29 Edwards, R (2000) Look before it leaps. *New Scientist* 167(No. 2244): 5.
30 ANZFA Draft Risk Analysis Report. Application A363. Food produced from glyphosate-tolerant canola line GT73. Australia New Zealand Food Authority, Canberra.
31 Eggum, BO (1981) Nutritional problems related to double low rapeseed in animal nutrition. In ES Bunting (ed.) *Production and Utilization of Protein in Oilseed Crops*. Martinus Nijhoff Publishers, The Hague, Boston, London, pp. 293–310.
32 Sorrells, ME (1999) Comparative genomics in crop improvement. *Proceedings 11th Australian Plant Breeding Conference*, CRC for Molecular Plant Breeding, Adelaide, Volume 1: 56–63.
33 Simmonds, DH (1989) *Wheat and Wheat Quality in Australia*. CSIRO, Australia.
34 Rahman, S, Li, Z, Batey, I, Cochrane, MP, Appels, R & Morell, M (2000) Genetic alteration of starch functionality in wheat. *Journal of Cereal Science* 31: 91–110.

7
INTELLECTUAL PROPERTY ISSUES

> So long as you're only quibbling about semantics,
> that's all right then
>
> Graeme Campbell[1]

CONTRADICTIONS

Intellectual property rights as applied to living organisms are 'not particularly intellectual'.[2] How can it be possible to patent a plant? The term 'patent' 'always implies a device or a structure or something that has been physically made, *de novo*, by someone who has thought up the idea or modified someone else's idea and has earned the right to the protection of a financial gain'.[3] A patent is 'a legal right which confers exclusive rights for an invention over a limited period'.[4] It is nonsense to claim that any extant form of life can be invented. Flowering plants have an independent evolutionary history, appearing on this planet more than 100 million years before humans did. No one can claim to have invented plants, yet patents are now being awarded for minute differences among cultivars of species that have been grown by our forebears for thousands of years. For humans to patent another living organism is both illogical and unethical, because it bestows a reward unfairly.

Why is it unfair? As I stated in my submission to the Australian House of Representatives Standing Committee on Industry, Science and Technology in 1991:

> Firstly, we inherit the base organism whose genotype is going to be modified. Perhaps it is only going to be modified in a single base in a

single gene somewhere in this complex genome. So although the person who is applying for a patent protection on doing this has, admittedly, made some intellectual or physical input into the process, he has inherited most of what he is then claiming to be protected by the patent. He is therefore deriving a benefit to which he is not properly entitled.[5]

Techniques are also being patented, regardless of whether they have sufficient novelty. Patents are supposedly given for 'what is new, and only to the first inventor'.[6] But this is not what has been happening. Most plant patents have not been judged properly for novelty. The current situation parallels the car manufacturing industry in the early years of the 20th century, just before Henry Ford successfully challenged George Selden's concocted patent for a horseless carriage.[7] Selden claimed that he had combined a number of elements into a new 'harmonious whole capable of results never before achieved'. But Selden's demonstration vehicle 'coughed into life, ran five yards, and then stopped dead'.[7]

Even though many utility patents in biotechnology have broad-ranging titles, every major multinational company involved in genetic modification has a patent on its own version of any fundamental technique that differs just a little from the generic. Another method of acquiring patents is to take over a company that has a desirable one, preferably so that litigation is curtailed or avoided (look again at Table 4.1). The number of patents held is proportional to the size of the company.

How is it possible to patent the base sequence of a piece of non-coding DNA? Finding out the base sequence is merely a discovery, and patenting a stretch of DNA, just in case it is useful in the future, makes nonsense of the terminology. This is exactly what Melbourne-based company Genetic Technologies has done.[8] But unless some genuine inventive purpose has been shown to depend on a particular base sequence, such patents are undeserved.

Patents are not the only concern here. What are Plant Breeders' Rights (PBR) and why do they take precedence over farmers' rights or the civil rights of an individual? Can PBR apply to genetically modified plants, or to plants that have only been selected, where there is no breeding program involved at all? Some argue that PBR cannot apply to 'a variety resulting from conventional cross breeding', and that patents should be sufficient to distinguish the products of genetic engineering.[9] Welcome to a world of confusion!

THE ORIGIN OF PATENTS FOR PLANTS

Ironically, patents for certain plants arose in the United States from some of the remarks made by that most prolific, generous and altruistic plant breeder, Luther Burbank (1849–1926). As quoted by Heitz,[10] Burbank observed:

> I despair of anything being done at present to secure to the plant breeder any adequate returns for his enormous outlays of energy and money. A man can patent a mousetrap or copyright a nasty song, but if he gives to the world a new fruit that will add millions to the value of the earth's annual harvests he will be fortunate if he is rewarded by so much as having his name connected with the result.

Luther Burbank was not complaining on his own behalf. He was looking to the future prospects of young people who might be deterred from embarking on a horticultural career. For his part, he was content:

> I do not envy any man living. I have never heard of any work or occupation or vocation that seems to me to rival that of the scientist, especially of the scientist who is equally a humanist and whose research and study and experiments and discoveries are all directed to the end that man may find this old sphere a better and more beautiful place in which to live.[11]

It was Luther Burbank's widow, Betty, who pursued the idea of rewards for plant breeders' inputs, and in 1930 the US Congress passed an amendment to the Patents Act that allowed some monetary reward to the breeders of fruit trees and ornamental plants. Exclusions at this stage included potato and Jerusalem artichoke.[10] To my mind it would have been more appropriate for the breeder's input to be rewarded with a royalty rather than a patent.

While still excluding open-pollinated varieties, the United States amended its patent law in 1953 to cover 'newly found plants in a cultivated state'.[4] That still sounds more like discovery than invention. Then patents on living organisms were consolidated with the *Diamond versus Chakravarty* case in 1980.[2,4,12] The US Supreme Court reversed the rejection of a patent application for a pseudomonad bacterium endowed with the ability to digest components of crude oil. Anan Mohan Chakravarty had transferred plasmids from three kinds of bacterium into a fourth. The Court ruled that:

> the patentee has produced a new bacterium with markedly different characteristics from any found in nature and one having the potential for significant utility. His discovery is not nature's handiwork, but his own; accordingly it is patentable subject matter.

Such a judgment was tailor-made for genetically modified plants, 'with markedly different characteristics from any found in nature'. Nevertheless, the United States 'is the only country in which patents for plant varieties *per se* have acquired some significance'.[10]

ALTERNATIVES TO PLANT PATENTS

Cultivated plants are called by many different common names, even within a single country. In contrast, a unique variety or cultivar name is needed if a variety is to be registered for commercial propagation and sale. Such registration can best be done outside the patent system. The word cultivar, often used interchangeably with variety, was coined from 'cultivated variety' by Liberty H Bailey in 1923, although credit is usually given to William T Stearn.[13] The concept is valid, although the abbreviation 'cv.' is not now placed before the cultivar name.

The origins of many cultivar names are lost in history. Today cultivar names are usually given by the breeder or originator. Just as in formal plant systematics, names can relate to some distinguishing feature of the variety, or they can be personal. A cultivar can be named for a person who:

- brought the variety to notice;
- grew the plant for a considerable period;
- was responsible for its selection or breeding;
- is being honoured by the originator.

To cover the commercialisation of cultivars produced by normal breeding procedures, various European associations of plant breeders negotiated the International Convention for the Protection of New Varieties of Plants. This convention was first adopted in 1961, and has been revised in 1972, 1978 and 1991.[10] The International Union for the Protection of New Varieties of Plants (UPOV) had 43 member states in mid-1999, with another 60 states having laws or draft laws based on one of the UPOV Conventions.[10] It is important to note that the 1991 version is the most severe, introducing restrictions on growers' traditional rights to save seeds (Chapter 8). The 1978 version is preferable.

In the United States, the Plant Variety Protection Act was introduced in 1970. This legislation sets out that a variety should be distinct, uniform and stable, criteria that are common in UPOV-compatible legislation around the world. A US cultivar protected by this Act has the letters PVP as a superscript after the name; for example, Super Sugar Mel^{PVP}, a dwarf snap pea resistant to powdery mildew. The duality of systems to protect intellectual property in plant varieties is ensconced in a provision of the World Trade Organization (WTO) concerning Trade-Related Aspects of Intellectual Property Rights (TRIPS). Members must agree to 'provide for the protection of plant varieties either by patents or by an effective *sui generis* system or by any combination thereof'. This simply means by an alternative system to patents 'of its own kind'.

REGISTRATION

To be in harmony with UPOV rules, the names given to plant cultivars that are to be propagated and sold through wholesale nurseries, retail nurseries or other retail outlets should not conflict with trade marks (see below) or trade names, nor be difficult, confusing or misleading in any way. A cultivar name must be unique within UPOV; it should be different from all pre-existing cultivar names registered for the same plant species or, in some cases, for groups of closely related species in different genera.

Proposals for new cultivars can be made by lodging descriptions and specimens with specialist registration authorities. In the United Kingdom, numerous specialist ornamental plant groups are associated with the Royal Horticultural Society (RHS), who organise comparative trials, the results of which are published in *The Garden*. Roses present special problems, as the unique cultivar name is actually the breeder's code name, and the same rose can have a variety of commercial synonyms in different languages and countries.[14,15] There are thousands of rose cultivars, and the best of the new releases in Australia are profiled in *Australian Horticulture*.[16,17] Potential new cultivars can be referred to a recently opened centre at the Adelaide Botanical Gardens.

Ornamental plants, such as roses and irises, must go through a trial process. A number of plants are grown in the trial beds for a specified period, then judged and ranked. Irises have become popular over the past 50 years. At first suitability for Australian climates was an all-important consideration. With these adaptations now well established,[18] iris cultivars are currently being assessed for the number of flowers produced in succession, as well as for the colours and qualities of the individual bloom.[19] Every year about 90 new irises are registered by the Australian Iris Society.[19]

Australian native plants are referred to the Australian Cultivar Registration Authority (ACRA), whose registrar, currently Iain Dawson, is based at the National Botanic Gardens, Canberra. ACRA is recognised internationally as the body responsible for the administration of the International Code of Nomenclature for Cultivated Plants as applied to Australian native species. New varieties of *Grevillea* and *Correa* are currently very popular. Overall, there are more than 300 cultivars of *Grevillea*. A project assisting ACRA to improve the documentation of Australian plant cultivars has been undertaken by the Royal Botanic Gardens, Melbourne, and the University of Melbourne's Burnley College.[20] The intention is to better describe past releases and so improve the determination of whether newly submitted specimens are really novel.

There is often no need to further protect ornamental cultivars by obtaining PBR (see below). This might sometimes be done for a rose or an Australian native plant, but many breeders have their own nurseries or established retail links, giving them what they believe is sufficient commercial advantage. There is no need in such instances to spend money on anything other than registration, which entails a modest fee (for example $50 for ACRA). Apart from the establishment costs, which might run into thousands of dollars, a PBR award costs $300 per year to maintain.

PLANT VARIETY LEGISLATION IN AUSTRALIA AND NEW ZEALAND

Both Australia and New Zealand passed Plant Variety Rights (PVR) legislation in 1987. Australia replaced the *Plant Varieties Rights Act* with the *Plant Breeders' Rights Act* in November 1994. The award of PBR is administered by a Registrar, currently Dr Doug Waterhouse, under the aegis of the Department of Agriculture, Fisheries and Forestry. The proceedings of the PBR office are recorded in the *Plant Varieties Journal*, which was first published in 1989. All varieties that existed at the time of the passing of the PVR Act are supposedly ineligible for rights, and are held to be varieties of common knowledge or belonging to the public domain.

The PBR system works best when genuinely novel varieties have been selected following deliberate hybridisation. This is the way the University of Sydney Plant Breeding Institute has proceeded in its breeding program for ornamentals, both for cut flowers and pot or bedding plants. Their range includes petunia, marguerite daisy (*Argyranthemum frutescens*), shasta daisy, kangaroo paw (*Anigozanthos*), Sturt's desert pea (*Clianthus formosus*), paper daisy (*Xerochrysum*, formerly *Bracteantha*), *Brachyscome* and *Diascia*. The marguerite daisies are known as the Federation Daisy series. Altogether 38 cultivars have been commercialised in 16 countries, and the Institute has 34 industry partners and agents. Between 1997 and 2000, annual sales increased from 1.9 million plants to 12 million, and the returns have helped support ongoing research and breeding programs.

The Australian PBR system has had problems adhering to the ideal that rights should be conferred only for cultivars that are distinct (that is, novel, uniform and stable) — criteria that must be met in order to comply with UPOV rules. Determining whether a variety is new has often presented a problem. It became evident in the mid-1990s that 'found' plants were being submitted for PBR. Australian native plants used as sources of food or flavour in the bush foods restaurant industry featured strongly, for example *Kunzea pomifera* Rivoli Bay, *Apium prostratum* Southern Ocean and *Mentha diemenica* Kosciusko, all identified as suspect in 1996.[21] Enquiries were ignored, and 'no evidence of any breeding programme for these "cultivars" has been forthcoming'.[21] The same can be said about more recent cases, such as the infamous Cardinal waratah (*Telopea speciosissima*), cultivated since 1955 but awarded PBR in 1997, and a rainforest fingerlime (*Citrus australasica*), now known as Rainforest Pearl.[22]

A propagation program is not a breeding program. The parent tree for this fingerlime 'cultivar' was described as 'a mature healthy tree 10–15 years old ... and it was raised from seed through two generations having originally been collected in the wild from State Forest approx. 30 years ago'.[23] The acid test for such plants is whether the alleged cultivars can be distinguished from plants still growing in the wild.

A critical survey of PBR awards encouraged by Rural Advancement Foundation International (RAFI) found 118 examples in which PBR should not have been awarded, an error rate of about six per cent of total applications.[24,25] Thirty-eight of these inappropriate awards concerned Australian native plants, the balance representing introduced species. Questions on notice to the Minister for Agriculture, Fisheries and Forestry were asked about the fingerlime award by Mr Michael Danby, MHR, member for Melbourne Ports, as follows:[23]

1. Is the Minister aware of the provisional PBR granted in January 1997 to a Ms Erica Birmingham of Bangalow in NSW for an Australian native fruit plant species *Citrus australasica* var. *sanguinea*: Rainforest Pink Pearls?

2. Does the Minister know if this plant variety is indeed a New and Distinct plant variety as required under the PBR Act?

3. Is the Minister aware that no description of this plant variety has ever appeared in the *Australian Variety Rights Journal* to substantiate the award of PBR to Ms Birmingham?

4. Is the Minister aware that this native plant is now being widely sold in nurseries across Australia under the protection of the PBR scheme without any justification of the grant or opportunity for the public to assess whether it is a new variety of plant, or to make comment or objection to this PBR grant?

5. Does the Minister know if this plant species was used and perhaps bred as a food plant by the local Aboriginal community of the area?

6. Will the Minister acknowledge that this Aboriginal community probably has traditional indigenous knowledge and thus intellectual property rights over varieties of this species of native citrus?

7. Will the Minister acknowledge that the PBR scheme as it currently operates does not prevent plant hunters from simply getting cuttings or seed of our indigenous plant species and then claiming intellectual property rights over such material?

8. What legislative or administrative action does the Minister intend to take so as to rectify this extraordinary situation?

Whatever answers have been recorded in *Hansard* (Australian Parliamentary Proceedings), nothing constructive has yet emerged by way of legislative amendment.

Varieties of common knowledge are supposed to be ineligible for the award of PBR. This has not prevented the PBR office making awards for such varieties, even after investigation periods of up to three years. Disease-resistant beans bred in Australia in the 1960s have been 'recycled' and the PBR awards given to a company in the Netherlands.[26] The alleged cultivars 'Simba' and 'Nelson' had their origin in Yanco, bred by Joe Sumeghy on behalf of NSW Agriculture, and accordingly they belong in the public domain.

Details of breeding programs are often sparse, the lack being disguised by the ambiguous term 'selection'. An assembly of plants transplanted from the wild into a garden bed can be described as a 'population',[27] but the word is not being used here in an ecological sense. Nevertheless, if a person makes a selection from such an artificial transplanted 'population' they can be regarded by the PBR office as the 'originator' of the variety.[27]

Even if genuine crosses have been performed, the progenitors are sometimes not clearly identified, and so they have not been included in trials that are supposed to demonstrate how distinctive or novel the new variety is. Sometimes 'new' varieties have simply been compared with one another, but not with their immediate progenitors, identified or not. In terms of scientific method, these comparisons are fundamentally flawed. Without comparisons with the progenitor(s), no scientific evidence of novelty has been advanced. PBR awarded in the absence of such published evidence are *ipso facto* invalid.

The stability of any claimed new variety is also a requirement of the PBR Act. This means not changing perceptibly from generation to generation. The submission of the proponent should be treated objectively, and set to one side. Stability should then be assessed independently of the proponent, and preferably by the PBR office. No one can assess stability without a 'snapshot' on at least two occasions, separated by an appropriate time interval. In New Zealand, trials are held in two consecutive years, and conducted by the awarding authority. This is better, but still not ideal. Five years is a more appropriate test interval. This would follow what is done by the RHS, which conducts trials and publishes the results widely. Trials of the same cultivars are repeated five years later, and the results are published again. Varieties that fail to perform consistently are identified, and those that perform as they did previously are endorsed.

A consequence of the UPOV requirement for stability is that F_1 hybrids should not be considered for PBR, except when such hybrids can be propagated asexually (vegetatively). F_1 hybrids reproducing sexually, whether by self-pollination or out-crossing, will segregate immediately. This instability is nothing new — it has been well documented since the rediscovery of Gregor Mendel's observations in 1900 (Chapter 1). What it means, logically, is that most F_1 hybrids would not be qualified to receive either cultivar names or PBR except by bending the stability rule within the Act itself.

THE STRANGE CASE OF A PEA CALLED TROUNCE

New Zealand's PVR system is different from the Australian PBR scheme in a number of respects, as mentioned with regard to their conducting trials twice and independently of the applicant. Nevertheless, the New Zealand system also gives undeserved protection to 'found' or 'essentially derived' varieties, as the example of Trounce illustrates.

The rights over a powdery mildew-resistant pea called Trounce were awarded first in New Zealand in 1992, and then in Australia.[28] These awards have looked puzzling from the outset. The applicants first appeared to be claiming that Trounce had originated as a mutation from Small Sieve Freezer, because the original plants were found growing in a crop of Small Sieve Freezer in Tasmania in 1987, and the origin was stated to be by 'field selection' (note the ambiguity). This would have been impossible from the number of differences Trounce showed among the parameters that were measured for the comparisons conducted in New Zealand in 1994 and then submitted in Australia (Table 7.1). Apart from resistance versus sensitivity to powdery mildew, there were differences in the number of leaflets per leaf at the first fertile node, the number of pods per plant, the number of seeds per plant, the mean seed mass and the extent of seed wrinkling, which was less in Trounce. The odds against this constellation of changes all happening at once are astronomic.

Table 7.1
Parameters published[28] for Trounce, Bounty and Small Sieve Freezer

Parameter	Trounce	Bounty	Small Sieve Freezer
Leaflets per leaf at first fertile node	5.90	5.60	4.10
Length of first pod (cm)	7.70	8.10	7.87
Width of first pod (cm)	1.33	1.28	1.33
Number of pods per plant	7.80	6.10	5.10
Number of seeds per pod	6.70	6.00	5.90
Number of seeds per plant	52.00	36.00	30.00
Mean seed mass (mg)	223.00	226.00	255.00

From information provided by the PBR office in 1999, it is clear that the applicants were aware that Trounce had come from a few stray seeds in a field of Small Sieve Freezer rather than being derived from this cultivar directly. They had described Small Sieve Freezer as the 'host', which is a misuse of a well-known biological term, more appropriate for relationships with pathogens or beneficial micro-organisms.

In summary, Trounce should not have been awarded PBR because:

- Trounce did not result from a breeding program.
- The applicants knew that Trounce was not the progeny of Small Sieve Freezer when they made their application.
- The progenitor of Trounce has not been satisfactorily identified, but it must have been a variety commercially available in 1987. Bounty is the prime candidate.

As to the source of Trounce, it is evident that the stray seeds already had a name, whether known to the applicants or not. This cultivar had to be one already in the public domain in 1987, and therefore not eligible to be nominated for PBR in its own right. By choosing Bounty as the only other variety in the comparative trials submitted with the Australian application, the applicants have thereby indicated that they believe Bounty to be the likely contaminant of Small Sieve Freezer seeds.

Strictly, however, Trounce's progenitor is unknown, and it follows that Trounce may not have been trialled against the originating material. So it is not possible to say in what respect it is different from its maternal parent or grandparents. Hence there is no evidence of novelty.

In response to the points made in summary above, Mr Bill Whitmore of the PVR Office in New Zealand defended the original award recognising Trounce as a distinct variety.[29] On point 1, he states 'Both the New Zealand Plant Variety Rights Act 1987 and the UPOV conventions recognise that varieties may originate from discoveries'. On point 2, he agrees. On point 3, he states that 'It is not necessary that the parentage of a variety be known in order for the variety to be eligible for plant variety protection.' As for deciding which varieties a 'novel' plant will be trialled against, he adds:

> even if there is a known parent we would not necessarily compare it with the new variety in trial. It is not uncommon for a new variety to be very distinct from its parent, in which case comparison in a growing trial is unnecessary. The varieties included with a candidate in a PVR comparative growing trial are normally the most similar ones. Only if the parent is one of the most similar varieties will it be included.

How does this apply in the case of Trounce? Apparently Bounty was included in the three-way comparison because it was 'one of the most similar ones'. But, despite Mr Whitmore's explanation of the rationale for trials, we know that Small Sieve Freezer is quite unlike Trounce (Table 7.1). Its inclusion was clearly just window dressing, with the effect of enhancing Trounce's apparent novelty.

If Trounce is indeed derived by selection from Bounty, is it sufficiently distinct to warrant the award of PBR? On the basis of the trial data supplied by the applicant, Trounce is higher yielding than Bounty. A strong difference in total number of pods per plant multiplied by a slight increase in the number of seeds per pod ensures many more seeds per plant for Trounce (Table 7.1). This looks like a valuable improvement. But in my experience such yield variation is within the bounds of that normally found for Bounty: between three and eight pods per plant for the first flowering. The size of the standard deviation (1.59) given in the *Plant Varieties Journal*[28] suggests a similar range for the Bounty that was included in this trial: adding and subtracting the standard deviation from the mean (a rough but useful guide) gives values of 7.69 and 4.51.

One would reasonably imagine that the onus of proof for claims supporting novelty would fall on the applicants for rights. This is not the case. Mr Bill Whitmore turns the onus of proof back on the people who object to the pseudo-scientific procedures that masquerade as varietal protection. He states 'I need more than speculation that the variety may be a derivative of "Bounty" or similar to a public domain variety in Australia. More specific and compelling evidence is needed to back a claim that "Trounce" should not have been protected in New Zealand.'[29]

This is amusing, considering that speculation is all the award was based upon. The objections are based on logical deduction. The time for 'compelling evidence' was when the application was first made. And in terms of the Australian regulations, Trounce should be considered as 'essentially derived' material in relation to its unidentified public domain progenitor. PBR are supposedly rejected in Australia for essentially derived varieties, but the New Zealand system refuses to rule on this matter.[30] Prior sale is also drawn in as an issue, as this supposedly represents another bar to PBR being awarded.[31]

It is also a condition of the award of PBR that a stock of the cultivar be kept in Australia and be made available to those with an interest in breeding. PBR do not preclude other parties using a protected cultivar in a breeding program. But approaches from Heritage Seed Curators Australia (HSCA) have met a stone wall. A. E. Stratton Crop & Food Research, the agent that supposedly had seed in Australia, did not have any seed of Trounce available. Because a stock of Trounce seeds was not held in Australia, the holders of the right are in breach of their award. No punitive action has been taken. The PBR office has also failed to retrial Trounce, after agreeing to do so, one reason being the unavailability of seed from this agent.

PROTECTING PUBLIC DOMAIN VARIETIES

What is worrying about the attitudes of both the New Zealand and Australian agencies dealing with PVR or PBR awards is that they seem unprepared to protect varieties that belong in the public domain. Found plants dressed up as discoveries are well represented among the many inappropriate PBR awards. It is left to interested citizens to see that varieties in the public domain are not 'discovered' by operators who seek to gain reward without effort. Alienation of our common genetic heritage must not be allowed to continue unchallenged. Every gardener who perpetuates plants can help to ensure that varieties belonging in the public domain remain there.

It is worse in Britain and Europe, because there it is now illegal to sell seeds of varieties that are no longer registered on the National List — a process that costs considerable money each year (currently 2000 pounds per cultivar). Some recent losses of cultivars in Britain have come from the inability of gardeners to afford the registration or even to know when a registration is likely to lapse. A correspondent in *The Garden* recently pointed out that a broad bean (*Vicia faba*) called The Sutton had lost its registration in the very year it had been given an Award of Gardening Merit by the RHS.[32] This thoroughly regressive and inequitable facet of loss adds to that from historical attrition.

One way around this problem for breeders of new vegetable varieties is that seedlings may be sold, whereas seeds cannot.[33] Medwyn Williams' new cultivars of leeks and celery, for example, can be purchased legally from him as seedlings (see Useful Addresses). Another way is to attempt to alter the legislation so that varieties of more interest to gardeners than to large-scale growers could be made exempt from registration fees.[34] The present system encourages the loss of genetic diversity among food and ornamental plants and removes all financial incentive for breeders to develop new varieties with such limited market appeal. The care of heritage varieties devolves to enthusiastic gardeners and seed-savers, as in Australia.[35]

ALTERNATIVES TO PBR AND PATENTS
TRADE MARKS

In accordance with the UPOV convention on cultivar names, a trade mark name should not be used as part of a varietal name. But confusion over trade marks is as widespread as confusion over PBR.[36] Trade marks are not supposed to be used as part of a cultivar name, but this happens all too often. There are two distinct problems. The first concerns sequence. Trade marks are being placed after the genus name, as in *Grevillea* Austraflora, or between species and cultivar names, as though they were part of the cultivar name. There are numerous examples in listings of new releases.[16,17,22,37,38] The second problem involves using the cultivar name as a trade mark by applying the superscript TM to it. The use of the initials TM or TP as a superscript for a trade mark is itself inadvisable, although frequently seen in *Australian*

Horticulture lists.[37,38] The superscript for a registered trade mark is R, usually encircled. It is also a condition of registration that the trade mark should not be used as part of a varietal name.

There are ongoing problems with the labelling of nursery plants, and not just because of confusion of trade marks with cultivar names. The legislation governing plant labels differs from state to state, and currently New South Wales, for example, lacks requirements as stringent as those found in states such as Victoria. The binomial species name (or genus name only if a hybrid) is a must for unequivocal identification, and should be in italics. The trade mark should be distinct, and perhaps preface all other names. Zee Sweet® peaches and nectarines and Nellie Kellie® passionfruit provide examples of clear labelling.

DEFENSIVE PUBLICATION

It is possible to ignore rights altogether and place the products of authentic breeding or selection in the public domain directly. Then it becomes important to back up such releases with published descriptions.[6] This precaution is advisable to protect these varieties from outright piracy — from having PBR or patents claimed by third parties who have simply grown the plants from seed once. This has already been attempted with the PBR system in Australia,[25] although following adverse publicity the applications made for two varieties of chickpea obtained from the collection at the International Crops Research Institute for the Semi-Arid Tropics (ICRISAT) were withdrawn by the applicant. Intellectual property claims cannot be made over any variety belonging to an 'in trust' collection, as this is expressly prohibited under an agreement with the Food and Agriculture Organization of the United Nations, which came into effect in December 1994.[39]

Registration of new cultivars of Australian plants with ACRA is itself a form of defensive publication. ACRA will include a specimen of the plant in its herbarium collection and add a description of the cultivar to its website (see Useful Addresses). Since all new PBR applications over Australian native plants will in future be referred to ACRA for an opinion, it becomes less likely that PBR will be given for an already registered cultivar.

My own new powdery mildew-resistant snow peas and semi-leafless garden peas have been placed in the public domain through seed-saving networks: the Seed Savers' Network (SSN) and HSCA. These varieties were bred by cross-pollination and selection, using Dwarf Oregon or Kodiak as sources of powdery mildew resistance, which is a recessive condition in peas. The program began in 1994 and the first descriptions were published five years later when seeds were released.[40] Seeds are available on request to any *bona fide* member of SSN or HSCA.

Neither defensive publication, nor releasing seeds, cuttings or plants to others, would disqualify a breeder from seeking PBR at any time in the future. This follows from the current interpretation of the PBR Act by the PBR Registrar, setting out his reasons for not revoking the PBR awarded for Cardinal waratah.[27] The PBR system stands in stark contrast to a patent system, where prior disclosure undermines the application for a patent and 'novelty' means 'that an invention has not previously been disclosed by publication or by use in the market'.[2]

CONCLUSION

Many in the horticultural and agricultural industries believe that we do need a system like PBR that permits exclusive rights to the propagation and sale of new cultivars. Without it, overseas interests might breed new cultivars of Australian plants and not distribute them back to Australia.[41] Others prefer to do without PBR, saving its costs, and relying on reputation and media exposure for peak sales. The horticultural industry is prone to fashions and trends, with peak sales for many cultivars lasting only a few years. Such varieties do not warrant protection for 20 years, and the considerable costs of PBR can well be avoided and added to the bottom line.

The Australian Plants Society and the Australian Flora Foundation support research aimed at bringing wild plants into cultivation as a conservation measure. But these associations do not support abuses of the registration or rights systems. Major flaws in the current PBR scheme include failure to properly safeguard public domain plants or cultivars, failure to revoke rights inappropriately given, and failure to acknowledge the intellectual property of the original custodians of plant genetic resources.[42] These difficulties have not been adequately addressed in recently proposed amendments to the PBR Act. Nevertheless, Section 6 of the Act permits genetically modified plants to receive PBR awards, and it is anticipated that transgenic plants will be protected in this way in Australia.

REFERENCES
1 Campbell, G (1991) Statement as a member of the House of Representatives Standing Committee on Industry, Science and Technology. 7 February, *Hansard*, pp. 822–23.
2 Blakeney, M (1999) Patented technologies. *Proceedings 11th Australian Plant Breeding Conference*, CRC for Molecular Plant Breeding, Adelaide, Volume 1: 13–16.
3 Murray, DR (1991) Submission to the House of Representatives Standing Committee on Industry, Science and Technology. 7 February, *Hansard*, p. 815.
4 Juma, C (1989) Life as intellectual property. In C Juma *The Gene Hunters — Biotechnology and the Scramble for Seeds*. Princeton University Press, Princeton, New Jersey, pp. 149–78.

5 Murray, DR (1992) In *Genetic Manipulation: The Threat or the Glory?* Report by the House of Representatives Standing Committee on Industry, Science and Technology, Canberra, p. 229.
6 Roberts, T (1994) Defensive publication. In The Crucible Group *People, Plants and Patents*. International Development Research Center, Ottawa, pp. 78–80.
7 Lacey, R (1987) *Ford — The Men and the Machine*. Pan Books, London.
8 Hope, D (2002) Gold in the DNA junk yard. *The Weekend Australian*, 16–17 February, p. 7.
9 Wong, SSH (1999) Protecting innovations in plant technology. *Proceedings 11th Australian Plant Breeding Conference*, CRC for Molecular Plant Breeding, Adelaide, Volume 2: 13.
10 Heitz, A (1999) Plant variety protection and cultivar names under the UPOV Convention. In S Andrews, AC Leslie and C Alexander (eds) *Taxonomy of Cultivated Plants: Third International Symposium*. Royal Botanic Gardens, Kew, UK, pp. 59–65.
11 Burbank, L & Hall, W (1927) *The Harvest of the Years*. Houghton Mifflin, Boston and New York.
12 Shiva, V (1997) *Biopiracy — The Plunder of Nature and Knowledge*. South End Press, Boston Massachusetts.
13 Anonymous (2001) William Thomas Stearn CBE, VMH. *The Garden* 126(7): 508.
14 Manners, MM (1999) Rose registration: Cultivar names, code names and selling names. In S Andrews, AC Leslie and C Alexander (eds) *Taxonomy of Cultivated Plants: Third International Symposium*. Royal Botanic Gardens, Kew, UK, pp. 117–24.
15 Spencer, R (2001) A rose by ... . *Australian Horticulture* 99(6): 38–40.
16 Thomas, S (2000) Roses 2000. *Australian Horticulture* 98(3): 44–66.
17 Thomas, S (2001) Roses — what's new in 2001. *Australian Horticulture* 99(3): 39–53.
18 Grosvenor, G (1985) Iris — elegant and irresistible. *Gardenscene* 1(4): 14–16.
19 Cattanach, D (ed.) (2000) *The Iris Society of Australia Yearbook 2000*. The Iris Society of Australia, Melbourne.
20 Spencer, R & Adler, M (2000) Students sort out Australia's cultivars. *Australian Horticulture* 98(2): 30–34.
21 Murray, DR (1997) Ecological and horticultural implications of the Australian native foods restaurant industry. *Native Plants for New South Wales* 32(2): 22.
22 Washington, S & Boucher, A (1999) New plant releases 1999. *Australian Horticulture* 97(7): 31–74.
23 Hankin, W (1999) The PBR grant for 'Rainforest Pink Pearls' Fingerlime. *The Curator* 13: 28.
24 Hankin, W (1998) Australia bungles 'Plant Breeders' Rights'. *The Curator* 10: 4–12.
25 Hankin, W (1999) The PBR situation in Australia. *Proceedings 11th Australian Plant Breeding Conference*, CRC for Molecular Plant Breeding, Adelaide, Volume 2: 4–9.
26 Murray, DR (1999) Breeding peas and beans. In David R Murray *Growing Peas and Beans*, Kangaroo Press, Sydney, pp. 46–51.
27 Waterhouse, D (2001) Letter to HSCA President, 18 April.
28 Anonymous (1997) Application 95/217. *Plant Varieties Journal* 10: 23.
29 Whitmore, W (1999) Letter to HSCA President Bill Hankin, 5 July.
30 Whitmore, W (1999) Plant Variety Rights in New Zealand. *Proceedings 11th Australian Plant Breeding Conference*, CRC for Molecular Plant Breeding, Adelaide, Volume 2: 10–11.
31 Stearne, PA (1999) Hot spots of legislative uncertainty/controversy in the PBR Act. *Proceedings 11th Australian Plant Breeding Conference*, CRC for Molecular Plant Breeding, Adelaide, Volume 2: 12.

32 Larkcom, J (1999) Vegetable madness. *The Garden* 124: 922.
33 Stebbings, G (2000) Keeping them crossed. *The Garden* 125: 824–27.
34 Cherfas, J (2000) Legalise diversity. *The Garden* 125: 212–13.
35 Fanton, M & Fanton, J (1993) *The Seed-Savers' Handbook*. The Seed-Savers' Network, Byron Bay, NSW.
36 Maxwell, JV & Vaughan, P (1999) Using trade marks for plant varieties. *Proceedings 11th Australian Plant Breeding Conference*, CRC for Molecular Plant Breeding, Adelaide, Volume 2: 14–15.
37 Thomas, S (2000) Plant releases for the year 2000. *Australian Horticulture* 98(7): 38–75.
38 Thomas, S (2001) Plant releases for the year 2001. *Australian Horticulture* 99(7): 31–66.
39 ETC Group (Action Group on Erosion, Technology and Concentration) (formerly Rural Advancement Foundation International) (2001) News release. 30 October.
40 Murray, DR (1999) Appendix 1 — Pea varieties. In DR Murray *Growing Peas and Beans*, Kangaroo Press, Sydney, pp. 69–75.
41 Berney, J (2000) WA efforts keep our waxflowers on top. *Australian Horticulture* 98(8): 14–19.
42 Simpson, T & Jackson, V (1998) Effective protection for indigenous cultural knowledge: A challenge for the next millennium. *Indigenous Affairs* 3: 44–56.

8
IMPACTS OF GENETICALLY MODIFIED PLANTS IN THE THIRD WORLD

> Agricultural practices that rely heavily on applications of expensive nitrogenous fertilisers and pesticides are no longer sustainable. There are some ecologically sound alternative approaches to solving agricultural problems, one of which is to breed plants with improved defences against a variety of pests and pathogens, and with better capacities to utilize nutrients economically, or withstand environmental stresses.
>
> David R Murray[1]

POPULATION SCENARIOS

The world's human population has recently passed six billion[2] and will continue to increase well into the 21st century, probably passing 10 billion before 2050. Two countries, India and the People's Republic of China, now have more than a billion people each. Globally, in 1998, there were five births for every two deaths.[2] Fertility rates are declining country by country, but it will be a considerable time before arrivals balance departures. How will so many people be fed and clothed? Some estimate that yields from existing agricultural lands will need to increase by 75 per cent by 2020,[3] and ultimately they will need to triple. Claims from multinational agribusiness companies that genetically modified plants will have a dominant role in feeding the world's multitudes have a superficial air of authenticity. It would be easy to be lulled into a false sense of security.

Before assessing what contribution might be expected from genetically modified plants and the companies promoting them, it is worthwhile questioning some of the assumptions behind these population projections. The amount of food produced per year at present, if replicated every year, might be sufficient for a long time to come if less were wasted or lost to post-harvest pests and diseases. Although the extent of such loss is difficult to estimate, ways of reducing waste can be relatively simple, such as providing shelters to protect against rain during harvest or building secure storage containers.[4] With encouragement, such as that provided by the United Nations Development Program (UNDP), more food could be produced closer to where it is needed in large cities, reducing the need for long distance transport and attendant spoilage.[5]

There also needs to be a more equitable distribution of food. Most malnourished people live in food-exporting nations, where growing the crops that earn foreign exchange takes an unwarranted priority over locally important crops and human wellbeing. Today Australia grows its own chickpeas, but the cultivated kabuli chickpea whose seed proteins I studied in the 1980s[6] was exported to Australia from Mexico! In Brazil, a major food exporter, 70 million people cannot afford to eat properly.[7] Some of the enormous quantities of wheat, maize and grain legumes presently used as animal food could feed humans directly, cutting out the inefficiency of consumption by intermediary birds and animals. Lot-fed beef cattle are the most wasteful, with an estimated eight units of grain needed to produce one unit of meat.[2]

Apart from these considerations, let us assume that the areas used for growing food plants remain much as they are at present, or decline even further. In this setting, the track record of genetically modified food crops can be seen to run contrary to what is required. Yields on an area basis have not increased for the major genetically modified crop plants — they have decreased, as evident for soybean and canola (Chapter 3). And regardless of how its yield compares with other rice cultivars, golden rice by itself cannot prevent malnutrition and starvation (Chapter 5). Where is there any evidence that transgenic plants could possibly double or treble yields per unit area?

A SECOND GREEN REVOLUTION?

Many are parroting Monsanto[7] and hailing the advent of genetically modified crop plants as a second Green Revolution. To judge whether this metaphor is apt, it is appropriate to review the achievements and shortcomings of the first Green Revolution. In the 1960s and 1970s, a number of new varieties of dwarf spring wheat were produced by Norman Borlaug and his colleagues at Centro Internacional de Mejoramiento de Maiz y Trigo (CIMMYT) in Mexico. High-yielding maize varieties were also developed there and, like the wheat cultivars, have been adopted in many developing nations. To produce the increased yields they did, these varieties were heavily dependent on synthetic fertilisers and pesticides.[8]

Similarly, new varieties of rice were bred at the International Rice Research Institute (IRRI) in the Philippines. They had shorter stalks, to prevent lodging (falling over under the weight of grain), and faster times to maturity, reduced from about 160 to 110 days.[9] Subject to adequate water supply, this allows three crops to be grown each year rather than one or two. From 1967 to 1992, estimated world rice production doubled. In Indonesia it tripled, from 15 to 48 million tons.[9]

But there were backward steps too. The yield increases were erratic, and sometimes there were huge unforeseen losses. Although resistant to many diseases, IRRI rice varieties are susceptible to ragged stunt and wilted stunt viruses, and sometimes to rice brown spot disease, caused by the fungus *Helminthosporium oryzae*. Insecticides have been used to kill the insects that spread the viruses, so that rice now ranks second only to cotton in total amounts of pesticide applied per crop. Because pesticides are not sufficiently selective, beneficial organisms, including frogs and 'complementary harvests'[7] such as fish, shrimps, crabs and native plants, are killed as well as pests.

The yield of a pesticide-dependent rice might be higher than the cultivars it is compared with, but this comes at the cost of the complementary harvests, which are not considered and which somehow have to be replaced. A simple comparison of rice yields forestalls a full appreciation of how productive rice-paddies can be when natural predators control pest insects and beneficial microbial associations in the soil and water are not annihilated by pesticide contamination. The adverse effects of pesticide exposure on human health[7] are considerable, but are also neglected. Rice yield is important, but it is not the only parameter we should be judging.

Another deleterious consequence of the Green Revolution has been the disappearance of traditional plant varieties, abandoned in the rush to grow seemingly more attractive releases. The traditional varieties are then unable to be found when sought after again for some worthwhile breeding quality. Fowler and Mooney have described how this happened with rice, millet, sorghum, okra, potato and other crops.[8] The example of potato is particularly telling. Dr Carlos Ochoa had collected 45 primitive potato varieties in two areas of northern Peru for use in his breeding program at the International Potato Center (CIP) in Lima, but on returning to these places 20 years later, in the early 1970s, he was unable to find most of them. Ironically, he attributed their loss to the popularity among growers of a cultivar that he himself had bred, named *Renacimiento* (rebirth).[8]

With promises of higher yields from more desirable varieties, many farmers have been beguiled into buying seeds they could not really afford. Selling seeds to farmers who then need to borrow to pay for the seeds and the necessary fertilisers and pesticides might be good for the companies involved, but it is economically and socially disruptive. A single crop failure is then enough to bring finances to a crisis, precipitating the loss of the land and the ability to grow even enough food for the farmer's family. Consequent migrations from rural areas to cities simply add to the unemployment–poverty cycle, while decreasing the number of skilled farmers. This pattern has been typical of countries participating in the Green Revolution. A second Green Revolution will certainly bring more upheaval and dislocation, with surviving farmers forbidden to save seeds for replanting. Seeds replete with Terminator technology would enforce the 'client' relationship between farmer and seed provider.

RISKS ASSOCIATED WITH GENETIC UNIFORMITY

Many writers have assumed that there is something intrinsically unnatural and undesirable about monocultures. A monoculture on a huge scale is inadvisable, as there is nothing to temper the extremes of sun, wind or rain, all of which can have devastating effects on yield and soil retention. Monotony — growing the same monoculture year after year — is also undesirable, as it encourages specific pests and diseases whose tenure is not disrupted by crop rotation. But monocultures on a moderate or small scale are useful, and for some plants it is difficult to conceive of any other practical method of growing and harvesting them. As I pointed out in 1984:

> It should be emphasized that it is not the crop monoculture itself that is undesirable, as evidenced by the capacity of wild cereals to form natural stands. Rather, it is the genetic constitution of any crop that will help or hinder it under stress.[6]

Reliance on potatoes with a narrow genetic base that lacked resistance to the late blight fungus (*Phytophthora infestans*) led Europe, Britain and Ireland to their disastrous potato famine from 1845 to 1850. By some quirk of fate, Australia had already received a blight-resistant variety, Early Manistee, grown first by convicts on the islands of Bass Strait, then later on the mainland.[10] The Reverend Chauncy Goodrich brought two blight-resistant varieties from Peru to the United States in about 1847: the Wild Peruvian and the Rough Purple Chilli.[11] He grew thousands of seedlings, finally selecting the Cuzco from the Wild Peruvian, and the Garnet Chilli from the Rough Purple Chilli. With these two varieties, especially the Garnet Chilli, he provided the foundation for modern potato breeding.[10] More than 500 blight-resistant cultivars trace their descent from the Goodrich selections.

A more recent instance of genetic uniformity proving a liability is the prevalence of type-T cytoplasm in maize, with its accompanying susceptibility to *Helminthosporium maydis* (Chapter 1). Most of the maize crop in the United States was wiped out by this fungus in 1970, the year in which Norman Borlaug accepted his Nobel Peace Prize. In Australia, the predominance of an aphid-prone lucerne cultivar, Hunter River, left farmers vulnerable to aphid infestations that reduced the lucerne crop by 60 per cent between 1979 and 1980.[12]

The uniformity of genetically engineered plant varieties exposes growers to a similar high risk of crop failure — if there is a genetic factor that leaves a crop vulnerable, then it is likely to be the same in every plant. This risk could be countered by increasing the numbers of cultivars sown, but little thought seems to have been given to the benefits of endowing a great many alternative cultivars with any particular engineered attribute. Roundup Ready soybean, for example, does not cope with heat stress as well as other varieties currently grown in the United States.[7] This is one reason for the lower yield displayed by the engineered variety (Chapter 3). Growers rushed to adopt the lower-yielding Roundup Ready variety only because of subsidies from the American government. According to Clare Murphy, 'Farmers are not interested in growing more competitive varieties when they can be subsidised for growing glyphosate-resistant soy beans.'[13]

Nevertheless, its recent fall in popularity with growers may be explained by heightened awareness of the greater risk of failure associated with this variety. With increasingly hot seasons, the possibility of a disastrous year for the Roundup Ready soybean is becoming more likely. It would have been prudent on the part of Monsanto to have planned for tougher successors to their prototype, with a view to its eventual replacement. How could a single modified cultivar possibly suit all growers in a country as diverse in climate and microclimate as the United States?

Soybean (*Glycine max*) arose in Asia, and hundreds of lowland and highland cultivars have been perpetuated by growers there for many centuries. A recent breeding program in Vietnam has utilised the genetic diversity represented by a collection of more than 600 cultivars to provide a dozen new cultivars with increased yields (1.5 to 2 tonnes per hectare compared with 1 tonne per hectare prior to 1998).[14] Other improvements contributing to these yield increases include disease resistance, wider tolerance of temperature extremes, and more compact production times. One early maturing summer variety (AK-02) fits neatly between rice crops, enabling a beneficial crop rotation.[14] With these new varieties, it is now possible to have three seasonal soybean crops each year (spring, summer and winter).

Premature flowering of short-day-sensitive soybean cultivars has been a major difficulty in Australia. This was overcome by conventional breeding: hybridisation, selection, multiple backcrossing and further selection.[15] Two new cultivars with flowering delayed by 12 to 16 days under spring conditions in Queensland display a consequent doubling of yield.[15] Cultivars like these show more potential for lifting soy production in the tropics and subtropics than any genetically modified variety publicised to date.

In contrast to the situation with Roundup Ready soybean, the production of more Bt-cotton varieties is very likely. There is an advantage in providing a plant with the capacity to produce multiple forms of the Bt protein (Chapter 6) and, at least in Australia, these genetic modifications have been introduced to a range of cotton cultivars by CSIRO Plant Industry.

SUSTAINABILITY AND FERTILISERS

Conventional sources of phosphorus-rich fertilisers are practically exhausted, and future requirements present a major problem, as discussed in Chapter 4. It is certainly possible that genetically modified legumes could make better use of phosphorus that is in the soil but locked up in unavailable forms. If shown to be safe, such genetically modified plants should be welcomed for the contribution they can make to lessening the need for a diminishing resource.

The supply of nitrogen is better assured. Although urea and ammonia can be manufactured, this is at a considerable cost in terms of energy and carbon dioxide production from fossil fuels. Nitrate manufacture from ammonia entails even higher inputs of fossil fuel energy and carbon dioxide outputs. In practice, fertilisers tend to be broadcast infrequently and in excessive amounts. This heavy-handed use of nitrogenous fertilisers has two prominent adverse effects: enriched runoff and consequent eutrophication of waterways, and, especially under waterlogged conditions, conversion of nitrate to nitrogen oxides[16] that escape to the atmosphere and contribute to the greenhouse effect. A third adverse effect occurs when plants store excessive nitrate in the parts that are harvested and eaten. Free nitrate in plant foods is potentially toxic, because it is converted to a carcinogen, nitrosamine, by intestinal bacteria.

The best way to increase the available nitrogen content of arable land is to grow legumes as part of a crop rotation. Legumes function as a gateway for the assimilation of atmospheric nitrogen by hosting nitrogen-fixing bacteria in the nodules on their roots. Some writers have only a vague idea of what transpires, believing that the bacteria provide 'various nitrogen compounds that can be used by the plant'[17] in exchange for sugar. In fact, these bacteria release ammonium ions to the plant, receiving in exchange a number of metabolites necessary for their own growth. The plants assimilate the ammonium ions with some of the proceeds of current photosynthesis, converting this simple form of nitrogen into a complete range of amino acids, proteins, cofactors, nucleic acids and pigments, eventually producing edible pods or protein-rich seeds. The shed parts of legumes, and their roots, provide assimilated forms of nitrogen to the soil, ultimately benefiting non-leguminous plants. Approximately 20 per cent of the nitrogen assimilated by a field of lupins, for example, is available to the next crop.[18]

Although peas and beans of several kinds are the most familiar legumes in Western countries,[19] a much wider variety is grown throughout Asia and Africa. In India 17 species of legume are commonly grown, although just five predominate: chickpea (*Cicer arietinum*), pigeon pea or red gram (*Cajanus cajan*), green gram or mung bean (*Vigna radiata*), black gram (*Vigna mungo*) and groundnut or peanut (*Arachis hypogaea*). All are grown as annuals, in rotation with millets or other cereals. In Yunnan province in southern China, the broad bean (*Vicia faba*) is grown in fields that will subsequently receive the rice crop. At maturity the whole plant is harvested and, after separation of the seeds, the tops are used to feed the water buffalo that help to plough the ricefields. Rotation with legumes and co-cultivation with the water-fern *Azolla*, which hosts the nitrogen-fixing cyanobacterium *Anabaena azollae*, provide low cost and environmentally benign ways of sustaining soil fertility without the application of synthetic nitrogenous fertilisers.

We know very little about soil micro-organisms. Some scientists believe that most species of soil bacteria have still not been described. Unknown relationships between plants and soil micro-organisms could sometimes explain the mysterious ability of 'economic' varieties of crop plant to resist disease and flourish with low fertiliser inputs. Recent research[20] reveals considerable cause for optimism about nitrogen fixation that directly benefits non-leguminous plants, especially monocotyledons. Selected varieties of tropical grasses such as elephant grass (*Pennisetum purpureum*) and species of *Brachiaria* grown in South America can gain up to 40 per cent of their nitrogen requirement from soil bacteria belonging to the genera *Herbaspirillum* and *Azospirillum*, respectively.[21] Sugar cane can acquire up to 60 per cent of its nitrogen requirement from nitrogen fixation by *Gluconacetobacter diazotrophicus*.[22] The amounts of applied nitrogenous fertilisers can be correspondingly reduced, provided that the soil is free of toxic substances and able to support these bacteria.

There are many other important growth-stimulating effects of soil bacteria,[23] not directly related to nitrogen fixation. Some bacteria convert elemental sulfur to sulfate ions, some release compounds that facilitate the uptake of iron (siderophores) or other minerals, and some are believed to achieve their beneficial effects by promoting the synthesis of plant growth hormones.[24,25,26] One of these interactions, between rice and a strain of *Rhizobium leguminosarum*, has evolved in the Nile delta in Egypt. Here the local rice has been cultivated in rotation with the usual host for this bacterium, berseem clover (*Trifolium alexandrium*), for more than 700 years.[26]

TRUE MEASURES OF PRODUCTIVITY

Growing a number of different plant species together provides an effective way of controlling pests and ameliorating environmental extremes. Stressed plants are unproductive and severely stressed plants will die. Under excessive heat, they close the pores (stomata) in their leaves to prevent water loss, thereby halting photosynthetic gas exchange and preventing fixation of carbon dioxide. But when crop plants are surrounded by other plants grown as windbreaks, shade or groundcover, these all help to minimise stress and so enhance productivity for the whole plant community. These additional plants can also provide edible parts such as fruits, or nectar and pollen for beneficial insects. A good example is the Sri Lankan forest garden system, which includes more than 20 multifunctional tree species.[27,28] As already mentioned in relation to rice cultivation, a more accurate measure of productivity requires a summation of all the items harvested from a system, not just the yields of the single crop under discussion.

Biological diversity promotes productivity in another way also. Growing more than just a single cultivar of any given crop is a hedge against the risk of total loss, and traditionally provides for sequential harvesting. A once only harvest can be totally lost if soaked by rain while drying out. Even with all going well, synchronised maturity can create a scarcity of labour and equipment. A series of smaller harvests spreads the risk of loss and provides continuous employment for more people.

The size of a farm can limit its productivity. The largest farms are not the most productive per unit area, essentially because they lack biodiversity and are more subject to extreme conditions. The examples from a United Nations survey instanced by Luke Anderson[7] suggest that an increase in productivity of up to three-fold might be expected from reducing farm sizes to the optimal, which in absolute terms will vary in different parts of the world.

'Organic' systems that avoid using herbicides and pesticides are demonstrably productive. Cuba was forced to become self-reliant for food production during the trade embargo imposed by the United States, and of necessity returned to labour-intensive low-input systems, with farm areas closer to the optimal.[7] More recently, Vietnam has had to recover from decades of conflict and rehabilitate the land necessary for cultivation. Age-old agricultural practices have been reinstated, encouraged by the Vietnamese Community Action Programme Against Hunger, Malnutrition and Environmental Destruction (VACVINA) and supported by the United Nations Educational, Scientific and Cultural Organization (UNESCO).[29] In the Red River

Delta, a major rice-growing area, every household has surrounding gardens where vegetables, herbs and fruits are grown together:

> Plants are grown in the garden in a system of tiered cultivation, in which various species are intercropped and overlapped to make full use of solar energy and soil nutrients. Fruit trees are interspersed with vegetables, beans and tuber crops, which grow in the shade. Other legumes are grown around the perimeter of the garden, and timber trees and rattans are planted to form the green fences.[29]

Other models have been developed that are more suitable for the Mekong Delta, the foothills, and the mountains. VACVINA has successfully promoted the 'nutrition plot', featuring amaranth (*Amaranthus gangeticus*), legumes, carrots, yams and sweet potatoes, and complemented by papaya (paw paw) and bananas. Adoption of VACVINA recommendations has led to a measurable improvement in nutrition (Table 8.1). By extrapolation, every nation could become self-reliant for basic food production by emulating Cuba and Vietnam.

Table 8.1
Improvements in diet resulting from the promotion of intensive small-scale farming in Vietnam[29] (grams per adult per day)

	Before 1985	Since 1985
Tubers	60.5	61.2
Beans	31.0	33.4
Oils/fats	4.0	8.9
Meat	18.6	43.4
Fish	—	2.0
Vegetables	253.7	373.0
Fruit	18.2	66.6
Sauce	11.9	7.4
Sugar	2.6	6.9
Energy equivalent (kcal)	2175	2478

SEED GUARDIANS

The practice of saving farm-grown seeds has been targeted by multinational seed producers so that they can increase their markets and returns. But provenance is all-important to minimising the risks of crop failure. Saving seeds for replanting is an integral part of the process whereby varieties adapted to particular sets of circumstances have been developed and perpetuated. Those plants best adapted to local conditions are the most likely to survive to maturity and give rise to the next generation. Over hundreds of years, deliberate selection of fruits and seeds from the healthiest plants has reinforced the association of characteristics that make each variety successful. As plants were spread geographically by human migration, or by exchange of seeds, diversification continued. In the case of rice, for example, 120 000 varieties have been recognised.[9]

These traditional ways of generating and conserving plant varieties are still enshrined in the cultural heritage of most nations. The responsibility for the seed supply can be mainly individual, or communal, where a village's supplies are placed under the care of designated 'seed guardians', as in Nepal.[30] Seed-saving can also be both individual and collective, as in China. Under the Chinese *baogan* policy, growers are expected to maintain their own seed lines, but there are back-up supplies of seeds to help those who have a shortage or a crop failure.[4]

In every country, the skills necessary for growing plants are no longer as highly regarded as they ought to be. Where local traditions have been eroded by mistaken ideas of progress and traditional varieties have been lost, assistance in reinstating former cultural practices, multiplying scarce local varieties, or providing replacements has sometimes been given by seed-saving volunteers. Since 1988, the founders of the Seed Savers' Network in Australia, Jude and Michel Fanton, have personally provided their expertise in Nepal,[31] India,[32,33,34] Cambodia,[32] Sabah (Malaysian Borneo),[35] Zimbabwe,[36,37,38] Kenya,[37] Cuba,[39,40] the Solomon Islands,[40,41] Fiji,[42] and Tonga.[43] As in Australia, their role is to co-ordinate local efforts to propagate and distribute suitable varieties of food plants and to promote awareness of the urgent need to rescue and perpetuate local varieties. They sometimes have to counter the influence of aid agencies such as USAid, who distribute commercial hybrid seeds and chemicals in a way that encourages farmers to turn away from their traditional varieties.[44] As Michel Fanton said of the situation in Nepal:

> Whatever comes from the outside world, especially in such a remote place, always seems better. The new experience confirmed my reticence to send seeds of foreign varieties of vegetables to projects as they could displace some endangered local varieties, especially when it comes to cross pollinators such as mustard, spinach, silver beets, radishes, other brassicas, and onions.[31]

Since October 1997, the Fantons have provided short courses twice a year to assist other volunteers, who in turn go and reinforce the local networks in such countries as the Solomon Islands,[36,40,41] Cambodia,[45] Malawi,[36] India,[45,46] Ecuador[45] and, most recently, East Timor.[47] Support for these volunteers has been enlisted from bodies such as the Australian Youth Ambassador for Development Program.

If this has not already been done, the volunteers write a manual on appropriate seed saving or propagation techniques, and preferably arrange a translation into the local language. The basis for these manuals is the book that Jude and Michel Fanton published in 1993,[48] with adaptations to local plants and circumstances. For example, saving seeds is particularly difficult in places where tropical humidity accelerates deterioration in seed viability. To counter this, seeds are often stored near cooking fires, which is not an ideal situation either. Sometimes seeds are stored in dried, smoked fruits, a customary procedure applied to eggplants in India,[49] and paralleling my own preferred method of storing tomato seeds in sun-dried fruits.[50]

India has been particularly vigorous in defence of its local varieties and of farmers' rights. Two prominent community groups have been formed, led by Dr Vandana Shiva and Dr Vanaja Ramprasad, respectively. Vandana Shiva is the Director of the Research Foundation for Science, Technology and Natural Resource Policy in New Delhi, and a well-known author. Vanaja Ramprasad, her former colleague, now heads the GREEN Foundation situated near Bangalore. Encouraging farmers to continue growing their traditional crops for local food supply, rather than switching to non-adapted export-oriented alternatives, is a top priority for both.

Vandana Shiva organised the First International Training Course on In Situ Conservation of Agricultural Biodiversity at Dehra Dun in 1997, at which many local delegates described their success in conserving varieties of rice and millets, including ragi or finger-millet (*Eleusine coracana*), fox tail millet (*Setaria italica*), bajra or pearl millet (*Pennisetum typhoides*) and little millet (*Panicum miliaceum*).[32,33] Apart from their obvious diversity and nutritional contributions, these millets are well suited to being grown under hot, dry conditions.[51] Vandana Shiva has since conducted a 'Monsanto Quit India' campaign,[34] adopting the Ghandi stance of peaceful resistance to injustice. More recently, the GREEN Foundation has organised a number of conferences, including a workshop titled 'Conserving Seed Diversity for Domestic Food Security', which facilitated the formation of the South Indian Seed Network in June 2001.[47]

TREATY NOW

The United Nations Food and Agriculture Organization Conference in Rome adopted an International Treaty on Plant Genetic Resources for Food and Agriculture on 3 November 2001.[52] Support for the Treaty was overwhelming, with 116 votes in favour, none against, and two significant abstentions: the United States and Japan. It comes into force once it has been ratified by 40 member states. The treaty will encourage Consultative Group on International Agricultural Research (CGIAR) institutes (such as IRRI, CIP and the International Crops Research Institute for the Semi-Arid Tropics) to continue their own plant breeding programs for the primary benefit of Third World nations, who are the donors of most of the varieties held in the institutes' collections. The breeding efforts of these institutions are complementary to grass roots efforts to protect the availability of locally adapted varieties, and more relevant to the real needs of Third World nations than the breeding goals of multinational companies. This treaty will also make it more difficult for unscrupulous seed recipients to pirate 'in trust' varieties (Chapter 7).

REFERENCES

1   Murray, DR (1991) Preface. In DR Murray (ed.) *Advanced Methods in Plant Breeding and Biotechnology*. CAB International, Oxford, pp. ix–x.
2   Reid, TR (1998) Feeding the planet. *National Geographic* 194(4): 56–75.
3   Rosielle, A (1999) Corporate view. *Proceedings 11th Australian Plant Breeding Conference*, CRC for Molecular Plant Breeding, Adelaide, Volume 1: 6.
4   Murray, DR (1990) Conclusions. In DR Murray *Biology of Food Irradiation*. Research Studies Press, Taunton, UK, pp. 217–24.
5   United Nations Development Program (UNDP) (1996) *Urban Horticulture — Food, Jobs and Sustainable Cities*. UNDP Publication Series for Habitat II, Volume 1, United Nations Plaza, New York.
6   Murray, DR (1984) The seed and survival. In DR Murray (ed.) *Seed Physiology. Volume 1. Development*. Academic Press, Sydney, pp. 1–40.
7   Anderson, L (2000) Genetic engineering and farming. In L Anderson *Genetic Engineering, Food, and our Environment*. Scribe Publications, Melbourne, pp. 55–82.
8   Fowler, C & Mooney, PR (1990) Genetic erosion: Losing diversity. In C Fowler and PR Mooney *Shattering — Food, Politics, and the Loss of Genetic Diversity*. Arizona University Press, Tucson, pp. 54–89.
9   White, PT (1994) Rice — the essential harvest. *National Geographic* 185(5): 48–79.
10 Murray, DR (2000) Root vegetables and tubers. In DR Murray *Successful Organic Gardening*. Kangaroo Press, pp. 67–71.
11 Webber, HJ & Bessey, EA (1899) Progress of Plant Breeding in the United States. *Yearbook for 1899*. USDA, Washington, DC, pp. 465–90.
12 Auricht, GC (1999) Lucerne — an Australian breeding success story. *Proceedings 11th Australian Plant Breeding Conference*, CRC for Molecular Plant Breeding, Adelaide, Volume 2: 71–76.
13 Murphy, C (2001) Travel report — WSSA Meeting and mid-west USA visit. *A Good Weed* 23: 6–8.
14 Long, TD & Thang, NQ (1999) Soybean breeding in Vietnam. *Proceedings 11th Australian Plant Breeding Conference*, CRC for Molecular Plant Breeding, Adelaide, Volume 2: 259–60.
15 James, AT (1999) Raising soybean yield and expanding production through application of crop physiology to agronomy and breeding. *Proceedings 11th Australian Plant Breeding Conference*, CRC for Molecular Plant Breeding, Adelaide, Volume 2: 27–28.
16 Davidson, BR & Davidson, HF (1993) Nitrogen — sources and cycling in Australia. In BR Davidson and HF Davidson *Legumes — the Australian Experience*. Research Studies Press, Taunton, UK, pp. 84–130.
17 Handreck, K (2001) Understanding fertilisers. In K Handreck *Gardening Down-Under: A Guide to Healthier Soils and Plants*, 2nd edn, Landlinks Press, Melbourne, pp. 115–36.
18 Davidson, BR & Davidson, HF (1993) Legume crops. In BR Davidson and HF Davidson *Legumes — The Australian Experience*. Research Studies Press, Taunton, UK, pp. 346–79.
19 Murray, DR (1999) *Growing Peas and Beans*. Kangaroo Press, Sydney.

20 Kennedy, IR (2001) Biofertilisers in action. *Australian Journal of Plant Physiology* 28: 825–27.
21 Reis, VM, dos Reis, FB Jr, Quesada, DM, de Oliveira, OCA, Alves, BJR, Urquiaga, S & Boddey, RM (2001) Biological nitrogen fixation associated with tropical pasture grasses. *Australian Journal of Plant Physiology* 28: 837–44.
22 Boddey, RM, Polidoro, JC, Resende, AS, Alves, BJR & Urquiaga, S (2001) Use of the ^{15}N natural abundance technique for the quantification of the contribution of N_2 fixation to sugar cane and other grasses. *Australian Journal of Plant Physiology* 28: 889–95.
23 Ryder, MH, Stephens, PM & Bowen, GD (eds) (1994) *Improving Plant Productivity with Rhizosphere Bacteria*. CSIRO, Melbourne.
24 Okon, Y & Labandera-Gonzalez, CA (1994) Agronomic applications of *Azospirillum*. In MH Ryder, PM Stephens and GD Bowen *Improving Plant Productivity with Rhizosphere Bacteria*. CSIRO, Melbourne, pp. 274–78.
25 Riggs, PJ, Chelius, MK, Iniguez, AL, Kaeppler, SM & Triplett, EW (2001) Enhanced maize productivity by inoculation with diazotrophic bacteria. *Australian Journal of Plant Physiology* 28: 829–36.
26 Perrine, FM, Prayitno, J, Weinman, JJ, Dazzo, FB & Rolfe, BG (2001) *Rhizobium* plasmids are involved in the inhibition or stimulation of rice growth and development. *Australian Journal of Plant Physiology* 28: 923–37.
27 Gunasena, HPM (ed.) (1993) *Multipurpose Tree Species in Sri Lanka*. University of Peradeniya, Sri Lanka.
28 Murray, DR (1997) The global carbon cycle and sustainable development. In DR Murray *Carbon Dioxide and Plant Responses*. Research Studies Press, Taunton, UK, pp. 213–16.
29 National Association of Vietnamese Gardeners (undated; post 1988) *VAC and VACVINA — Vietnamese Community Action Programme Against Hunger, Malnutrition and Environmental Destruction*. Published with the assistance of UNICEF in Vietnam.
30 Fanton, M (2001) personal communication, 13 October.
31 Fanton, M (1999) Life and seeds in Nepali Highlands. *Seed Savers' Network Newsletter* 26: 8–9.
32 Fanton, J & Fanton, M (1997) Teaching seed saving overseas. *Seed Savers' Network Newsletter* 23: 2.
33 Fanton, J & Fanton, M (1998) International biopiracy. *Seed Savers' Network Newsletter* 24: 6.
34 Fanton, M (1999) India leads resistance to genetic piracy. *Seed Savers' Network Newsletter* 26: 4.
35 Fanton, J & Fanton, M (1998) Replanting the forest in Borneo. *Seed Savers' Network Newsletter* 25: 1.
36 Fanton, J & Fanton, M (1998) Positive solutions. *Seed Savers' Network Newsletter*, No. 25: 1.
37 Fanton, J & Fanton, M (1999) Seed savers in Africa. *Seed Savers' Network Newsletter* 26: 11.
38 Fanton, J & Fanton, M (2000) Seed security for Africa. *Seed Savers' Network Newsletter* 28: 6.
39 Fanton, J & Fanton, M (1996) Seed savers concept exported to Havana. *Seed Savers' Network Newsletter* 20: 3.
40 Fanton, J & Fanton, M (2000) International seed projects. *Seed Savers' Network Newsletter* 28: 7.
41 Fanton, J & Fanton, M (1999) Solomon Islands network set to expand. *Seed Savers' Network Newsletter* 26: 10.
42 Fanton, J & Fanton, M (1997) Pacific seed conference. *Seed Savers' Network Newsletter* 22: 1.

43 Fanton, J & Fanton, M (1996) Tongan Save Our Seed project. *Seed Savers' Network Newsletter* 21: 2.
44 Fanton, M (1999) Nepali non-violent resistance. *Seed Savers' Network Newsletter* 26: 5.
45 Fanton, J & Fanton, M (2000) Trainees go off. *Seed Savers' Network Newsletter* 28: 3.
46 Fanton, J (2001) South Indian seed network formed. *Seed Savers' Network Newsletter* 31: 23.
47 Fanton, J & Fanton, M (2000) Report from East Timor. *Seed Savers' Network Newsletter* 29: 1,25–26.
48 Fanton, M & Fanton J (1993) *The Seed Savers' Handbook for Australia and New Zealand*. The Seed Savers' Network, Byron Bay, NSW.
49 Ramprasad, V (2000) personal communication, 11 November.
50 Murray, DR (2000) Flowers, fruits, seeds and seedlings. In David R Murray *Successful Organic Gardening*. Kangaroo Press, Sydney, pp. 43–47.
51 Simmonds, NW (ed.) (1976) *Evolution of Crop Plants*. Longman, London and New York.
52 Action Group on Erosion, Technology and Concentration News — Seed Treaty Approved in Rome. 6 November 2001.

9
LOOSE ENDS

> ... it is not expected that there will be any wholesale movement towards registration and sale of genetically-modified foodstuffs. No food company wishes to be first with a new technology whose acceptance by consumers is doubtful.
>
> Australian Science and Technology Council (1993)[1]

RECAPITULATION

Genetically modified plants have been in the offing since the 1970s. The debate about their limitations is an old one. Campaigns and demonstrations against them, however, are a recent phenomenon, prompted by the predominance of just a few varieties, such as Monsanto's Roundup Ready soybean, which accounted for 90 per cent of the soybean crop planted in the United States at its peak, and Bt-transformed maize or cotton (INGARD). These varieties all have major defects, and yet their releases were approved. Crop plants resistant to systemic herbicides such as glyphosate or glufosinate will incorporate herbicide into the edible parts of the plant, substantially increasing herbicide contents compared with products from plants grown with pre-emergent herbicides only (Chapter 3). Potentially toxic bacterial proteins under the ubiquitous viral promoter CaMV35S are produced in all parts of the plant, so that Bt-transformed maize produces uncontrollable clouds of pollen containing Bt-protein, and grain products that may not be suitable for human consumption (Chapter 6).

These same plants were approved for release in Australia by the

Genetic Manipulation Advisory Committee (GMAC) and the Interim Office of the Gene Technology Regulator (IOGTR). Field trials were also approved for genetically modified canola of various kinds, including those expressing aspects of Terminator technology (Chapters 2, 4). As noted previously, canola is a cultivated plant that is also a weed (Chapter 3). Despite claims from Aventis and others that containment measures were being complied with in their field trials, very clearly they were not.[2] Workers hand-harvesting seeds from mature plants did not use protective clothing, and were carriers of seeds lodged in their ordinary clothing, especially their shoes. Stock animals were allowed to eat plant residues. Volunteer seedlings continued to sprout after the conclusion of the trials, as farmers were not fully informed of the nature of the plants trialled and the extra precautions necessary to prevent their spread.

The breaches of field containment measures that occurred near Mt Gambier came to light through the observations of local resident Leila Huebner,[2] not through inspections by the regulators. There were breaches of containment guidelines in Tasmania also, at 11 out of 58 sites operated by Aventis and Monsanto.[3] The spread of genes from transgenic canola to other cultivars in Tasmania represents a major hazard for organic canola farmers, whose markets are now in jeopardy unless they can secure new sources of seed from unmodified varieties. The placement of so many trial plots in sensitive areas was ill-advised and unnecessary, and could be interpreted as deliberate sabotage. In the same vein, 'accidental' contamination of canola seeds sold in Britain and Europe[4] with seeds of a genetically modified variety defies Hanlon's Razor (Never attribute to malice that which is adequately explained by incompetence).

NEW LEGISLATION FOR REGULATION

How will the development and release of transgenic plants be governed in the future? Is there any sign that anything has been learned from the mistakes of the past? From 21 June 2001 the provisions of the *Gene Technology Act 2000* have come into operation. Growing genetically modified plants must be licensed by the Office of the Gene Technology Regulator (OGTR), under the Federal Department of Health and Ageing, and the Minister. The Gene Technology Regulator is Dr Sue Meek. She has access to advice from three new committees: the Gene Technology Community Consultative Committee, the Gene Technology Ethics Committee, and the Gene Technology Technical Advisory Committee. This last committee replaces the Genetic Manipulation Advisory Committee (GMAC) and will have 20 part-time members, representing a much wider range of disciplines than GMAC did. The other two new committees have

12 part-time members each, and are in place to transmit community and ethical concerns. Anyone with appropriate qualifications and experience can be nominated for Ministerial appointment to any of the three committees, and for the first round of appointments, there were ten volunteers for each available position.

At the developmental stages, there is a requirement for 'contained facilities'. The standard for physical containment is generally PC2. In addition to basic safety precautions, PC2 requires procedures for decontamination and safe disposal of wastes that present a biohazard. Institutional Biosafety Committees must be formed, and these are to be responsible for implementing standards and ensuring that facilities and projects are registered with the OGTR.[5] In the recent past, the CSIRO Plant Industry Institutional Biosafety Committee has not only regulated CSIRO's projects, but has also advised GMAC about projects occurring elsewhere. In the 1980s, adherence to these standards was fairly haphazard,[6] but there is now much better compliance with containment guidelines, at least at the stages of development that occur before field trials.

Another change following from the *Gene Technology Act 2000* is that the locations of approved field trials are to be made public on a website[7] unless this information is regarded as commercial in confidence. This is a welcome change in principle, although one might predict the outcome. However, the OGTR is empowered to inspect sites, report breaches and impose fines — powers that were lacking with GMAC's voluntary guidelines. It is anticipated that only 20 per cent of total sites could be inspected in any one year,[2] a far cry from what is really necessary. Vigilance on the part of interested members of the public is still warranted, and there is a continuing need to support the GeneEthics networks (see Useful Addresses). States and municipalities will be able to declare 'GM-free' zones, and some have already done so.[8]

The total approval structure for proposals to grow genetically modified crop plants does not just involve the OGTR and its advisory committees. The National Registration Authority for Agricultural and Veterinary Chemicals has to approve the use of particular herbicides or pesticides. Food Standards Australia and New Zealand (FSANZ, which replaced the Australia and New Zealand Food Authority, ANZFA, in 2002) must also approve any plant products that are to be eaten. To date, ANZFA has not insisted on independent tests of food safety, but instead has relied on approvals from overseas when making their 'full assessment'. This rubber-stamping exercise will no doubt continue. No committee seems prepared to consider the unique hazards posed by unexpected genetic effects, and either the short term or long term effects of consuming bacterial and viral gene products on human health.

All of this means that it will be easier for applicants seeking release of genetically modified plants to gain approvals, because no single authority is taking an integrated view of the consequences of release and then consumption of products. Fragmented risk assessment like this will continue to overlook the obvious defects in released plants (Chapter 6). The Australian Conservation Foundation proposed a 'one stop shop' alternative,[9] but regrettably this is not the structure that has been adopted.

AVOIDING PRODUCTS OF GENETICALLY MODIFIED PLANTS

Consumer attitudes to genetically modified plants around the world have hardened, with Australia and the United States now catching up to Europe and Japan (82 per cent opposed).[10]

Despite industry opposition to labelling of food containing ingredients produced from genetically modified plants, this is mandatory from the end of 2002. There are many exceptions, however, and the threshold of one per cent leaves plenty of scope for skulduggery and carelessness. Processed foodstuffs lacking any significant nucleic acid or protein content, such as oils, starches or sugars, are exempt. Food processing agents such as lecithin, a phospholipid fraction often used as an emulsifier, are also exempt. Meat or milk produced from animals that have eaten genetically modified plant foods will not be labelled to indicate this. Similarly, seeds harvested from legumes hosting genetically modified *Rhizobium* bacteria in their root nodules will not be labelled to this effect. Restaurant meals or takeaway food will not need to admit to any transgenic sources.

Labels and advertisements can proclaim freedom from genetically modified ingredients. Checking compliance with the one per cent threshold will require analysts specialising in the use of gene probes, nucleic acid sequencing, protein identification and quantification. Companies such as GeneScan Australia Pty Ltd can provide this service (see Useful Addresses). Consumers will need to pressure the Federal Government to ensure that independent analyses are carried out often enough to instil confidence that food producers are actually fulfilling their obligations and labelling accurately.

The trustworthiness of labels is another matter. The country of origin of specific ingredients is concealed at present when foodstuffs are combined from different sources. 'Made in Australia from Local and Imported Ingredients' is uninformative and misleading. The use of the words 'Australia' or 'Aussie' in brand names is often just another marketing ploy, designed to deceive. To be certain of avoiding processed soy foods derived from imported soybeans, with their elevated content of glyphosate (Chapter 3), ensure first of all that the Australian manufacturer has undertaken to avoid genetically modified ingredients. Rather than trust the American 'Identity Preservation' system for separating unmodified soybeans, look for soybeans grown in Australia. As indicated in Chapter 8, new varieties suitable for spring-sowing under Australian conditions have been bred by conventional means and released by CSIRO Tropical Agriculture.[11]

Alternatively, consider replacing some or all of the soy in your diet. This might not be practicable if lactose-intolerance is the reason for consuming soy, but there is no category of essential nutrients or beneficial plant products that is unique to soybeans. For superior sources of essential amino acids, look to the many alternatives among other legumes, such as chickpeas,[12] lentils, peas and beans.[13] For phyto-oestrogens, there are alternative plant sources. Soy is highly promoted[14] but dangerous for people with some medical conditions, such as underactive (or surgically removed) thyroid[15] and hypertension.

Advice that 'eliminating all potential or possible sources of GM food products and ingredients from our daily food intake would result in the higher risk of an unbalanced and nutritionally barren diet'[16] is alarmist and unsound. At this stage such replacement concerns only processed foods — turning to minimally processed or unprocessed alternatives, free of added salt, oil or sugar, would provide health benefits that are self evident.

Beware of a host of products that might contain imported corn kernels, cornflour or corn syrup. The Bt-protein in StarLink corn, CryIXC, can cause allergenic reactions in a significant number of individuals (Chapter 6). Therefore it is possible that any other Bt-protein produced in maize kernels could cause similar problems. There is a cloud over genetically modified maize varieties of any kind (which additional herbicide would you like?), and avoidance is the safest policy. This will require reading labels very closely, as cornflour ends up in some surprising places, such as canned sardines with tomato sauce and packet icing mixture.

Bt-proteins in potatoes have not been cleared of causing serious ailments in rats (Chapter 6). Whenever possible, imported potato products should be identified as to country of origin and avoided if they come from countries that permit the growing of genetically modified varieties for human consumption.

AVOIDING PESTICIDE RESIDUES

Remember that the good oil is olive oil. Whatever improvements may be made to the composition of cottonseed oil in the future (Chapter 5), at present all cotton products come tainted with pesticide residues. Cotton is the most generously endowed recipient of pesticides of any crop.[17] Australian meat producers who fed their animals 'cotton trash' now know this to be true, and mushroom growers using 'cotton hulls and meal' instead of horse manure in their base medium have provided yet another way of contaminating foods with pesticide residues and cadmium.[18]

Even if pesticide application has been reduced by 50 per cent, as generally claimed for Bt-modified cotton, this is 50 per cent of a very large quantity in the first place, which the non-transformed cotton plants are still subjected to. Moreover, pesticides applied to Bt-cotton are concentrated in the second half of the growing season (Chapter 6). Any hydrophobic pesticide residues on or in the plants are likely to dissolve in the oil when the seeds are pressed. Cottonseed oil might not be fit for human consumption — strong evidence will need to be adduced that it is. Margarine manufacturers need to be quizzed about their possible use of cottonseed oil under the category 'vegetable oils'. Avoid margarines with non-explicit labels, and avoid deep-fried fast food. That is sound advice in any case.

A survey of Sydney market produce gives a good idea of which fresh foods are likely to have the greatest pesticide contamination and which are likely to have least.[19] The insecticide most frequently found above its maximum recommended level is endosulfan, although restrictions on its use have tightened since this survey was completed. Collectively, however, fungicides are the most frequently encountered contaminant. Those from the dithiocarbamate group occurred in 48.5 per cent of all 1566 samples, and the maximum recommended level was exceeded in 9.9 per cent of samples. These average figures do not give the complete picture, as many kinds of fruit and vegetables in the survey made a negligible contribution to this result. The products with the highest frequency of dithiocarbamate contamination among the samples taken included kiwifruit, chinese cabbage, capsicum, mangoes and strawberries. In some of these instances, most samples exceeded the maximum recommended level of dithiocarbamate.

As a general rule, it is best to avoid market produce at the beginning or the end of its season, when prices are highest and the temptation to sell before withholding periods have expired is very strong.

AVOIDING TOXIN-CONTAMINATED FOODS

Foods imported from other countries can pose significant extra hazards apart from whether they are derived from genetically modified plants. A recent example concerns products of capsicums infected with *Aspergillus* fungus, which produces significant aflatoxin contamination. Aflatoxins cause liver damage, including cancer.[20] According to Andreas Klieber of the University of Adelaide, 80 per cent of imported capsicum products have higher than the maximum permitted content of aflatoxins, especially dried and powdered products.[21] Yet capsicums are not difficult to grow in the home garden, without using any pesticides.[22,23]

SUPPORTING ORGANIC PRODUCTION SYSTEMS

The only sure way to avoid ingesting pesticide residues is to eat food from plants grown in the complete absence of pesticides, in uncontaminated soil. Avoidance of synthetic pesticides is a fundamental premise of any organic growing system. Other worthwhile features include the avoidance of expensive manufactured fertilisers, incorporating legumes in crop rotations (Chapter 8), rebuilding soil fertility without damaging the environment somewhere else,[18] and saving seeds for replanting. A number of different organic codes provide certified standards, which have been simplified by the National Association for Sustainable Agriculture Australia (NASAA) to two: 'organic' and 'in transition'. The term 'organic' now has connotations of health — for the individual and for the environment.

Organic Vita Brits are back on the shelf, as Goodman-Fielder (Uncle Toby's) have again been able to source wheat grown organically. The price difference between organic and non-organic Vita Brits is modest. Organic rice, however, presents a major problem. This has nothing to do with the way rice is grown, but with the way rice is marketed. In the Murrumbidgee Irrigation Area, rice has been grown for about 70 years. In that time, 15 cold-tolerant japonica varieties have been bred that are suited to the local growing conditions.[24] Yields of the order of nine tonnes per hectare are currently obtained, which is an extremely good figure, albeit at the expense of very high water consumption. The usual rotation is two seasons of rice, followed by one of wheat, barley or oats, then by one of legume pasture. This system is readily adapted to becoming organic, by dispensing with herbicides and pesticides and reducing the volume of water consumed.

However, organic rice growers are not free to process and sell their own grain. They are obliged to deal with the Rice Marketing Board,[25] which cannot guarantee freedom from contamination with non-organically grown rice. To avoid the likelihood of mixing, and so ensure fidelity to label, it is essential that the monopoly powers of the Rice Marketing Board be removed. A similar measure has been sought by growers of durum (pasta) wheat, whose export permits were previously delayed or blocked by the Australian Wheat Board. The Australian Competition and Consumer Commission (ACCC) should break up all grain monopolies so that there is a clear stream of organic produce, handled at different locations to non-organic, and certified by NASAA. There is no chance of keeping genetically modified grain products identifiable if we cannot first guarantee that any grain product labelled organic is indeed organic.

The safest food is the food that you grow yourself. Most people do not have the time or the space to be totally self-sufficient, but it is usually possible to grow something — a lemon tree plus some herbs in pots make a good start. Parsley, rosemary, oregano, garlic chives and mint can be available all year round, with others like basil and dill plentiful in season. As organic produce is often justifiably expensive because of the extra care taken in its production, you will score on freshness, safety and economy every time you grow something edible in your own garden.[26] There is plenty of help available from the seed-saving networks (see Useful Addresses), and it costs very little to belong.

REFERENCES

1. Australian Science and Technology Council (1993) *Gene Technology — Issues for Australia*. Occasional Paper No. 25, AGPS, Canberra.
2. SBS Television (2001) 'Insight' — Secret seeds, 21 June.
3. Phelps, B (2001) GE field trials. *The Seed Savers' Network Newsletter* 30: 12.
4. Tata, P (2001) Cotton-pickin' farce. *New Scientist* 172(No. 2316): 19.
5. Richardson, AE (2001) Containment procedures for products of gene technology: An Australian perspective. CSIRO Plant Industry, Canberra, <www.pi.csiro.au/briefings/issues/process.htm>, accessed 3 May 2001.
6. Murray, DR (1991) Submission to the House of Representatives Standing Committee on Science, Technology and Industry, 7 February, *Hansard*, pp. 811-13.
7. Office of the Gene Technology Regulator. Department of Health and Ageing, Canberra, <www.ogtr.gov.au>, accessed 2002.
8. Phelps, B (2000) Genetic engineering — freeze it for five years. *Habitat* 28(6): 15-22.
9. Phelps, B (1999) Say 'no' to gene tech's bitter harvest! *Habitat* 27(3): 13-20.
10. Angus Reid Survey (2000).
11. James, AT (1999) Raising soybean yield and expanding production through application of crop physiology to agronomy and breeding. *Proceedings 11th Australian Plant Breeding Conference*, CRC for Molecular Plant Breeding, Adelaide, Volume 2: 27-28.
12. Murray, DR & Roxburgh, CMcC (1984) Amino acid composition of the seed albumins from chickpea. *Journal of the Science of Food and Agriculture* 35: 893-96.
13. Murray, DR (1999) Eating peas and beans. In DR Murray *Growing Peas and Beans*. Kangaroo Press, Sydney, pp. 59-65.
14. Stock, S (2001) Women 'waste money' on soy help. *The Weekend Australian*, 25-26 August, p. 7.
15. Bye, C (2001) The overactive cancer. *Sun-Herald*, 26 August, pp. 12-13.
16. Duncanson, K (2000) Futuristic foods. *YourHealth* 1: 24-25.
17. Gatehouse, JA (1991) Breeding for resistance to insects. In DR Murray (ed.) *Advanced Methods in Plant Breeding and Biotechnology*. CAB International, Oxford, pp. 250-76.
18. Murray, DR (2000) Preparing the soil. In David R Murray *Successful Organic Gardening*. Kangaroo Press, pp. 33-39.
19. Plowman, T, Ahmad, N & Bower, C (1998) *Monitoring Pesticide and Cadmium Residues in Fresh Fruit and Vegetables 1992-5*. Horticultural Research and Development Corporation and NSW Agriculture, Orange.
20. Murray, DR (1990) Production of microbial toxins in stored irradiated foods and loss of quality in wheat. In DR Murray *Biology of Food Irradiation*. Research Studies Press, Taunton, UK, pp. 113-35.
21. Klieber, A (2000) *Chilli Spice Production in Australia*. Rural Industries Research and Development Corporation (RIRDC) Publication No. 00/33, RIRDC, Canberra.

22 Bailes, M (1999) *The Fragrant Chilli*. Kangaroo Press, Sydney.
23 Murray, DR (2000) Fruits as vegetables. In David R Murray *Successful Organic Gardening*. Kangaroo Press, pp. 75–80.
24 Ginnis, L (1999) Getting to the grain. *Geo* 53: 96–117.
25 ABC Radio (2000) 'The Country Hour'. 25 October.
26 Murray, DR (2000) *Successful Organic Gardening*. Kangaroo Press, Sydney.

USEFUL ADDRESSES

ACA — Australian Consumers' Association
57 Carrington Rd, Marrickville NSW 2204
phone (02) 9577 3399; fax (02) 9577 3377
ausconsumer@choice.com.au
www.choice.com.au

ACRA — Australian Cultivar Registration Authority
Australian National Botanic Gardens
GPO Box 1777, Canberra ACT 2601
phone (02) 6250 9472; fax (02) 6250 9474
www.anbg.gov.au/acra/acra-list-2000.html

Biotechnology Australia
www.biotechnology.gov.au

CSIRO Plant Industry
GPO Box 1600, Canberra ACT 2601
phone (02) 6246 4911; fax (02) 6246 5000
info@pi.csiro.au
www.pi.csiro.au

FSANZ — Food Standards Australia and New Zealand
(replaced ANZFA in 2002)
PO Box 7186, Canberra Mail Centre ACT 2610
phone (02) 6271 2222; fax (02) 6271 2278
slo@foodstandards.gov.au
www.foodstandards.gov.au

GeneEthics Network (Australian Conservation Foundation)
340 Gore St, Fitzroy VIC 3065
phone 1800 332 510 or (03) 9416 2222; fax (03) 9416 0767
geneethics@acfonline.org.au
www.geneethics.org

GeneEthics Sydney (Friends of the Earth)
PO Box 890, Crows Nest NSW 1585
geneethicssydney@hotmail.com
www.sydney.foe.org.au/gene_ethics

GeneScan Australia Pty Ltd
Suite 19, Technology Enterprise Centre, 2 Park Drive, Bundoora VIC 3083
phone (03) 9479 5055; fax (03) 9479 5056
info@genescan.com.au
www.genescan.com.au

GeneWatch UK
The Mill House, Manchester Road, Tideswell, Buxton SK17 8NY, UK
mail@genewatch.org
www.genewatch.org

GRAIN — Genetic Resources Action International
Girona 25, pral., E-08010 Barcelona, Spain
grain@grain.org
www.grain.org

GREEN Foundation
PO Box 7651, Bangalore 560 076, India
phone (080) 6097393; fax (080) 6651729
van@vsnl.com

ISIS — Institute of Science in Society
PO Box 32097, London NW1 OXR, UK
www.i-sis.org

NASAA — National Association for Sustainable Agriculture Australia Ltd
PO Box 768, Stirling SA 5152
phone (08) 8370 8455; fax (08) 8370 8381
enquiries@nasaa.com.au
www.nasaa.com.au

National Registration Authority for Agricultural and
Veterinary Chemicals
PO Box E240, Kingston ACT 2604
phone (02) 6272 5158
nra.contact@nra.gov.au
www.nra.gov.au

Natural Law Party, Wessex
nlpwessex@bigfoot.com
www.btinternet.com/~nlpwessex

New Scientist
www.newscientist.com

OGTR — Office of the Gene Technology Regulator
PO Box 100, Woden ACT 2606
phone 1800 181 030; fax (02) 6271 4202
ogtr@health.gov.au
www.ogtr.gov.au

Organic Herb Growers of Australia Inc.
PO Box 6171, South Lismore NSW 2480
phone (02) 6622 0100; fax (02) 6622 0900
admin@organicherbs.org
www.organicherbs.org

Organic Federation of Australia
PO Box Q455, QVB Post Office, Sydney NSW 1230
phone (02) 9299 8016; fax (02) 9299 0189
info@ofa.org.au
www.ofa.org.au

ORGAA — Organic Retailers and Growers Association of Australia
Box 12852, A'Beckett St Post Office, Melbourne VIC 3000
phone 1800 356 299 or (03) 9737 9799; fax (02) 9737 9499
oas@alphalink.com.au
www.lexicon.net/zebra/temp/robjordan

Organic Vignerons Association of Australia
PO Box 503, Nuriootpa SA 5355
phone (08) 8562 2122; fax (08) 8562 3034

Plant Breeder's Rights Australia
GPO Box 858, Canberra ACT 2601
phone (02) 6272 4228; fax (02) 6272 3650
pbr@affa.gov.au
www.affa.gov.au/pbr

Programme for Traditional Resource Rights
Oxford Centre for the Environment, Ethics and Society
Mansfield College, Oxford University, Oxford OX1 3TF, UK
users.ox.ac.uk/~wgtrr

DuPont Protein Technologies
PO Box 88940, St Louis MO, 63188, USA
www.protein.com

The Ram's Horn (Brewster Kneen)
S6, C27, RR.1
Sorrento BC, V0E 2WO, Canada
ramshorn@ramshorn.bc.ca
www.ramshorn.bc.ca

ETC Group (Action Group on Erosion, Technology and Concentration)
formerly RAFI — Rural Advancement Foundation International
Suite 200, 478 Rive Ave, Winnipeg MB, R3L 0C8, Canada
etc@etc.org
www.etcgroup.org

SSN — Seed Savers' Network
PO Box 975, Byron Bay NSW 2481
phone or fax (02) 6685 6624
info@seedsavers.net
www.seedsavers.net

Union of Concerned Scientists
2 Brattle Square, Cambridge MA, 02238-9105, USA
ucs@ucsusa.org
www.ucsusa.org/agriculture

USDA — United States Department of Agriculture
Animal and Plant Health Inspection Service
www.aphis.usda.gov/ppq/biotech

Williams, Medwyn
Llanor, Old School Lane, Llanfairpwllgwyngyll, Anglesey LL61 5RZ, UK
phone (01248) 714 851
medwyn@llanor.fsnet.co.uk
www.medwyns-prize-show-vegetables.com

GLOSSARY

acre	the imperial unit of area, equal to 4840 square yards, and approximately 0.4 hectare.
alleles	alternative forms of a gene. For a recessive condition to be expressed, two copies of the recessive allele must be present, one on each homologous chromosome. If a condition is dominant, a single copy of its allele is sufficient for expression.
amino acids	simple organic acids with a carboxylic acid group, an amino (NH_2-) group, and a side-chain. Twenty common amino acids form polypeptides. Essential amino acids are those that adult humans cannot synthesise and need to take in from dietary sources: methionine and cysteine (sulfur-containing amino acids), phenylalanine and tyrosine (aromatic amino acids), valine, leucine and isoleucine (branched chain amino acids), threonine, lysine and tryptophan.
amylase	an enzyme that can break down starch.
anther	the part of a stamen that produces the pollen grains, which contain the male gametes.
bacteriophage	a virus that attacks a bacterium.
bioaccumulation	the increase in concentration of a pesticide or herbicide as organisms are eaten by others along a food chain. For example, DDT applied to kill insects accumulated in birds, causing very thin eggshells and low reproductive success. Scavengers are usually the worst affected by bioaccumulation. Failure to consider bioaccumulation is a common fault of those releasing novel xenobiotic compounds for agricultural purposes, only to have them recalled with the benefit of hindsight.
binomial	the two-part scientific or systematic name of a species; the first name is the genus, the second the 'epithet'. A Latin form is used even if the words have Greek or other derivation, and by convention the binomial is italicised.
biotechnology	any technique used in the management of living organisms for

human purposes, such as using yeasts to produce beer or wine, or inoculating rice paddies with dried mats of blue-green algae. This term may now have a narrower focus, implying a modern or laboratory-based technique.

cDNA — complementary DNA, made from a messenger RNA template by reverse transcription. This form of nucleic acid would be chosen for gene transfer from a eukaryotic organism t o a bacterium, as prokaryotes lack the ability to process eukaryotic messenger RNA by excising intron sequences not present in the final transcript.

chlorophyll — a green pigment with a structure like haem, but with a central magnesium atom surrounded by a modified porphyrin ring. Chlorophyll plays a vital role in the chemical reactions of photosynthesis.

clone — plants arising from vegetative cells or protoplasts (that is, by mitosis) are genetically identical to the parent, and together with the parent represent a clone. In tissue culture, cells are not always identical, and somaclonal variation (Chapter 2) is observed. The term 'clone' has been misapplied to animal and human reproduction where the nucleus of an egg cell is replaced by the nucleus of a body cell from the male parent. Because the mitochondrial genome is maternal, the resulting embryo has received genetic contributions from both parents. Such an embryo is therefore not a clone of either parent. A plant derived in this way might be termed a cybrid (q.v.).

cloning — in gene technology, making many copies of a fragment of an organism's genome, after it has been spliced into a bacterial plasmid or lambda phage, or by polymerase chain reaction.

cotyledon — the first formed leaf of a plant embryo. Plants with one cotyledon are monocotyledons; plants with two are dicotyledons.

cultivar — a cultivated variety recognised as distinct from its wild progenitors and from other cultivars. Cultivar names should not be given to varieties that lack sufficient distinction.

C-value — the amount of DNA in a haploid nucleus, expressed in picograms (10^{-12} g). This measure is an indicator of the size of the nuclear genome, which has increased as flowering plants have evolved.

cybrid — a plant derived from a cell containing the nucleus from one species and the cytoplasm from another species.

cytochromes — haem pigments present in chloroplasts and mitochondria. They have a central iron atom surrounded by a porphyrin ring.

cytoplasm — a living plant cell protoplast consists of a nucleus, vacuole and cytoplasm, which is everything apart from these two entities. Cytoplasm includes the obvious organelles, such as microbodies, Golgi bodies, mitochondria, chloroplasts or related plastids, and the system of organised membranes called the endoplasmic reticulum. The soluble phase of cytoplasm is distinguished as the 'cytosol'.

denitrification — the reduction of nitrate ions to oxides of nitrogen or molecular nitrogen (N_2), carried out by certain aerobic soil bacteria, especially in the absence of sufficient oxygen under waterlogged conditions.

embryogenesis — cells belonging to normal (vegetative) cells of a plant, or protoplasts derived from such cells, can sometimes be induced to

	divide and differentiate into the parts of a new embryo. This process involves transfers through appropriate culture media.
endosperm	a distinct storage organ or tissue abundant in the seeds of many flowering plants, including cereals and *Solanum*, but absent or much reduced in others, such as peas and beans. In such instances, the embryo is adapted to accumulate reserves of starch, protein, minerals and oils. Endosperm cells are polyploid: triploid (3n) or greater.
enzyme	a catalyst of biological origin, usually a protein with the ability to catalyse a chemical reaction important to the metabolism of living cells. Enzymes speed up or facilitate reactions that would not otherwise take place under normal conditions of temperature and pressure. Most enzymes can be isolated and used in vitro. The common name of an enzyme says what it does, for example citrate synthase catalyses the synthesis of citrate (citric acid).
eukaryotic	in cells that are eukaryotic the nucleus is surrounded by a membrane, and there are other distinct membrane-bounded organelles with specialised tasks, such as microbodies, mitochondria and chloroplasts.
fertilization	the union of an egg cell with one of the sperm cells admitted to the ovule via a pollen tube, so forming the first cell of a new embryo (zygote). In flowering plants (Angiosperms), two sperm cells are produced within each pollen grain, and double fertilization is said to occur because the second sperm cell fuses with the two 'polar' nuclei to form the first endosperm cell. Although fertilization was first described in flowering plants in the 1840s, double fertilization was not discovered until 1900.
fluorescence	the emission of light of a different wavelength to light that has been absorbed.
gametes	a general term encompassing egg and sperm cells.
gene	a discrete unit of inheritance represented by a length of DNA located in a chromosome.
glycoprotein	a protein with glycosyl (sugar or saccharide) attachments, by linkage to the side chains of the amino acids asparagine, threonine, or serine.
greenhouse effect	the warming of the atmosphere due to the trapping of infrared radiation by gases such as carbon dioxide, methane and oxides of nitrogen. Increasing concentrations of carbon dioxide have two further effects: narrowing the pores in plant leaves, thus decreasing transpiration and promoting high temperature stress; and nutritional erosion: decreasing the contents of sulfur-enriched proteins and vitamins in seeds of major crop plants.
hectare	the metric unit of area, equal to 10 000 square metres, or 2.471 acres.
hybrid	the progeny of two distinct parents, which may be different species belonging to the same genus or two varieties of the same species. In rare cases, hybrids are formed across a generic barrier, for example *Chamelaucium floriferum* and *Verticordia plumosa* (Myrtaceae). First filial or F_1 hybrids are unstable because they are heterozygous for many genes, and alternative alleles will segregate whenever the second (F_2) and subsequent generations are produced sexually.

internode	the portion of a stem between two nodes, or points of leaf attachment.
meiosis	reduction division, during which the chromosomes are first separated into two homologous sets. Generally a diploid mother cell gives rise to four haploid cells, which include the gametes or their precursors.
mitosis	cell division that yields two cells similar to the cell that has divided; if a diploid cell undertakes mitosis, then normally the two products will also be diploid.
mutagenesis	the natural low incidence of mutations can be magnified by exposing seeds or other plant parts to ultraviolet or gamma radiation, or to chemicals such as bisulfite and ethylmethanesulfonate.
mutation	a change in a gene. The substitution of a single base in DNA can lead to a simple or 'point' mutation, changing the identity of a single amino acid in a particular position in a polypeptide. Deletion or insertion of one or more bases causes a 'reading frame' shift during translation, resulting in substantial changes to the amino acid sequence in a polypeptide. Even a point mutation can lead to the production of an inactive enzyme if it occurs in a critical position that contributes to the catalytic site. A good example is the abrupt conversion of coloured petals to white in many flowering plants, which indicates the loss of an active enzyme somewhere in the pathway leading to anthocyanin pigment production.
nitrification	the conversion of ammonium ions (NH_4^+) to nitrite (NO_2^-) and thence to nitrate (NO_3^-) by soil bacteria, especially *Nitrosomonas* and *Nitrobacter*, respectively. Most plants prefer to take up nitrate as nitrogen source.
nucleotide	a purine or pyrimidine base linked to a ribose or deoxyribose sugar, and one or more linked phosphate groups. Nucleotides act as precursors of DNA or RNA, and also participate in biosynthesis as substrates, or as precursors of cofactors necessary for certain enzymes to function.
organelles	*see* eukaryotic.
phloem	one of the two major vascular systems of plants, localised in veins and responsible for the transport of sucrose, amino acids, organic acids, minerals and water in a bi-directional fashion (upwards and downwards). Phloem transport cells are thin-walled and living, unlike mature xylem vessels, and can be destroyed by elevated temperatures. Viruses can be transported in the phloem if injected by an aphid.
plant breeding	the crossing of distinct parents followed by selection from the progeny constitutes breeding; selection alone does not constitute a breeding program. To recognise that crossing has taken place via natural agencies (wind or insect pollination) is valuable, but varieties so obtained are merely discoveries. A plant breeder worthy of the title will replicate a suspected parentage by experimental crossing and selection. An accurate definition of plant breeding is missing from the *Plant Breeders' Rights Act 1994*.
plasmids	small circular DNA molecules external to the bacterial chromosome; plasmids can be exchanged by bacterial cells, and this assists the spread of genes conferring antibiotic resistance.

polypeptide	a polymer comprising amino acids (and two imino acids, proline and hydroxyproline) joined by peptide bonds. The sequence of amino acids in a polypeptide is governed by the base sequence of the corresponding gene.
precautionary principle	if there is reason to believe that some human activities are dangerous to the environment or human health, then it is sensible to refrain from these activities even though there might be a lack of scientific information that would enable a faster decision one way or the other. A lack of scientific information should not be used as an excuse to continue with the status quo. For example, we now risk irreparable damage because various governments have dithered over carbon dioxide emissions.
prokaryotic	in prokaryotic cells the nucleus is not surrounded by a membrane, and is termed a 'nucleoid'. There are no distinct organelles like those that occur in eukaryotic cells. Bacteria and so-called 'blue-green' algae (cyanobacteria or Cyanophyceae) possess prokaryotic cells.
protein	one or more associated polypeptide chains.
recombination	during meiosis, corresponding lengths of some homologous chromosomes can be swapped so that the chromosomes then allotted to gametes are not identical with the chromosomes that were received from the parents. The enzymes that perform 'cutting and pasting' like this are the same enzymes that splice DNA to permit the incorporation of transgene packages.
ribozyme	by analogy with 'enzyme', an RNA molecule with the ability to catalyse cleavage of chemical bonds within the same RNA molecule, or in others.
seed	a fully developed ovule, containing a plant embryo, with or without distinct endosperm, and surrounded by a seedcoat or seedcoats. Sometimes fruit coats enclose the seed, as in the grains of cereals.
species	a 'kind' of living organism. The individual members of a species are interfertile (at least in theory, because this is prevented by self-fertilization in some plants, such as peas). A plant species may or may not have an impediment to breeding freely with its closest relatives in the same genus.
sterility	the inability to produce functional gametes, either egg cells, or sperm cells in pollen grains. Male sterility can be cytoplasmic or nuclear in origin; cytoplasmic male sterility involves a factor coded in the mitochondrial genome.
stigma	the receptive surface of the female part of a flower, which permits the germination of compatible pollen grains and supports the initial growth of pollen tubes as they proceed to the ovules.
symbiosis	a relationship between two contrasting organisms showing interdependence rather than exploitation, such as insects taking nectar and/or pollen from flowers, but bringing about pollination; or nitrogen-fixing bacteria living in the security of nodules on roots, exchanging ammonium ions for metabolites and shelter.
synergism	an interaction whereby the effect of two (or more) substances together is greater than the sum of the effects of the same concentration of each substance acting alone.
systemic	entering the vascular system and being transported throughout the plant.

taxonomy	the classification of living organisms.
transcription	the synthesis of messenger RNA (mRNA), which involves a gene in a strand of DNA acting as a template (Chapter 1). In eukaryotic organisms, polymerisation is followed by excision of introns (unexpressed intervening sequences) to yield one or more types of processed mRNA molecules used in translation (q.v.).
transformation	the stable introduction of DNA from an external source to a recipient genome.
transgenic plant	a plant that has been transformed by incorporation of a construct that includes a gene or genes from another species of plant, or any other kind of living organism, or a virus.
translation	the assembly of polypeptides following the migration of messenger RNA molecules to the ribosomes. The processes of transcription and translation allow a gene to specify the sequence of amino acids in a polypeptide, and hence the shapes and properties of proteins.
variety	a distinct type within a species, equivalent to cultivar for cultivated plants.
virus	a small, non-living entity that infects a target organism, taking over the synthetic machinery of each cell, and manufacturing many copies of itself. Symptoms of viral infection often include leaf 'mosaic' patterns, and tissue distortions. Viruses consist of a nucleic acid core surrounded by coat proteins. Plant viruses mostly have a core of RNA, but two groups possess DNA, which makes them potentially useful for genetic transformation of plants.
yield penalty	as applied to herbicide-resistant canola, or to genetically modified crop plants of any kind, the extent to which the yield falls short of that obtained from an unmodified cultivar growing under the same environmental conditions.
xenobiotic	as applied to synthetic chemical compounds, those that are so alien and novel that living organisms may have no previous exposure to anything like them in the course of their evolution, or if such exposure occurred, it was so long ago that the enzymatic capacity to modify such compounds has been lost. They often successfully challenge and overcome our detoxifying enzyme systems and immunological defences, so causing cancers.
xylem	one of the two major vascular systems of plants, localised in veins, and responsible for the transport of water, mineral ions and amino acids in an upwards direction, in response to the transpiration pull arising from the evaporation of water from pores in leaves. Xylem vessels or members are thickened, lignified, and dead at maturity, forming strong hollow tubes with capillary dimensions. Water and solutes move by cohesion (sticking together) and adhesion (sticking to the sides of the walls). Major solutes are transferred from the xylem to the phloem at night, when water movement has practically ceased.

FURTHER READING

Ableman, M (1993) *From the Good Earth — Traditional Farming Methods in a New Age*. Thames and Hudson, London.
Ackhurst, RJ (ed.) (1994) *Proceedings of the 2nd Canberra Meeting on Bacillus thuringiensis*. CSIRO Division of Entomology, Canberra.
Anderson, L (2000) *Genetic Engineering, Food, and our Environment*. Scribe Publications, Melbourne.
Andrews, S, Leslie, AC & Alexander, C (eds) (1999) *Taxonomy of Cultivated Plants: Third International Symposium*. Royal Botanic Gardens, Kew, UK.
Auld, BA & Medd, RW (1992) *Weeds — An Illustrated Botanical Guide to the Weeds of Australia*. Inkata Press, Melbourne, Sydney.
Balfour, EB (1975) *The Living Soil and the Haughley Experiment*. Faber and Faber, London.
Beder, S (1997) *Global Spin — The Corporate Assault on Environmentalism*. Scribe Publications, Melbourne.
Carr, DJ & Carr, SGM (1981) *Plants and Man in Australia*. Academic Press, Sydney.
Carson, R (1962) *Silent Spring*. Houghton Mifflin, Boston.
The Crucible Group (1994) *People, Plants and Patents*. International Development Research Centre, Ottawa, Canada.
Engel, K-H, Takeoka, GR & Teranishi, R (eds) (1995) *Genetically Modified Foods: Safety Issues*. ACS Symposium Series 65, Washington DC.
Fanton, M & Fanton, J (1993) *The Seed-Savers' Handbook*. The Seed-Savers' Network, Byron Bay, NSW.
Ford-Llloyd, B & Jackson, M (1986) *Plant Genetic Resources: An Introduction to their Conservation and Use*. Edward Arnold, London.
Fowler, C & Mooney, P (1990) *Shattering — Food, Politics, and the Loss of Genetic Diversity*. The University of Arizona Press, Tucson.
Glick, BR & Thompson, JE (1993) *Methods in Plant Molecular Biology and Biotechnology*. CRC Press, Boca Raton, Florida.
Graham, Frank Jr (1972) *Since Silent Spring*. Pan/Ballantine, London.
Hindmarsh, R, Lawrence, G & Norton, J (eds) (1998) *Altered Genes — Reconstructing Nature: The Debate*. Allen & Unwin, Sydney.
Harlan, JR (1992) *Crops and Man*, 2nd edn. Crop Science Society of America and American Society of Agronomy, Madison, Wisconsin.
Harlan, JR (1995) *The Living Fields*. Cambridge University Press, Cambridge.

Hoe, M-W (1998) *Genetic Engineering — Dream or Nightmare?* Gateway Books, Bath, UK.
Howard, A (1945) *Farming and Gardening for Health or Disease.* Faber and Faber, London.
Juma, C (1989) *The Gene Hunters — Biotechnology and the Scramble for Seeds.* Princeton University Press, Princeton, New Jersey.
Kyle, MM (ed.) (1993) *Resistance to Viral Diseases of Vegetables: Genetics and Breeding.* Timber Press, Portland, Oregon.
LeBarron, H & Gressel, J (eds) (1982) *Herbicide Resistance in Plants.* John Wiley & Sons, New York.
Liebman, M, Mohler, CL & Staver, CP (2001) *Ecological Management of Agricultural Weeds.* Cambridge University Press, Cambridge.
Masefield, GB, Wallis, M, Harrison, SG & Nicholson, BE (1985) *The Illustrated Book of Food Plants.* Peerage Books, London.
McLeod, J (1994) *Heritage Gardening.* Simon & Schuster Australia, Sydney.
Morrow, R (1993) *Earth User's Guide to Permaculture.* Kangaroo Press, Kenthurst, NSW.
Murray, DR (1988) *Nutrition of the Angiosperm Embryo.* Research Studies Press, Taunton, UK.
Murray, DR (ed.) (1991) *Advanced Methods in Plant Breeding and Biotechnology,* CAB International, Oxford.
Murray, DR (1997) *Carbon Dioxide and Plant Responses.* Research Studies Press, Taunton, UK.
Murray, DR (1999) *Growing Peas and Beans.* Kangaroo Press, Sydney.
Murray, DR (2000) *Successful Organic Gardening.* Kangaroo Press, Sydney.
Perring, FH & Mellanby, K (eds) (1977) *Ecological Effects of Pesticides.* Academic Press, London.
Peters, JA (ed.) (1959) *Classic Papers in Genetics.* Prentice-Hall, Englewood Cliffs, New Jersey.
Piperno, D & Pearsall, D (1998) *The Origins of Agriculture in the Lowlands Neotropics.* Academic Press, London.
Plowman, T, Ahmad, N & Bower, C (1998) *Monitoring Pesticide and Cadmium Residues in Fresh Fruit and Vegetables 1992–5.* Horticultural Research and Development Corporation and NSW Agriculture, Orange, NSW.
Posey, D & Dutfield, G (1996) *Beyond Intellectual Property: Toward Traditional Resource Rights for Indigenous and Local Communities.* IDRC, Ottawa, Canada.
Pretty, JN (1995) *Regenerating Agriculture.* Earthscan, London.
Simmonds, NW (ed.) (1976) *Evolution of Crop Plants.* Longman, London and New York.
Shewry, PR, Napier, JA & Davis, P (eds) (1998) *Engineering Crop Plants for Industrial End Uses.* Portland Press, London.
Shiva, V (1997) *Monocultures of the Mind,* 3rd edn. Third World Network, Penang.
Shiva, V (1997) *Biopiracy — The Plunder of Nature and Knowledge.* South End Press, Boston.
Shiva, V (2000) *Stolen Harvest — The Hijacking of the Global Food Supply.* South End Press, Cambridge, Massachusetts.
Suzuki, D & Dressel, H (1999) *Naked Ape to Superspecies.* Allen & Unwin, Sydney.
Suzuki, D & Knudtson, P (1989) *Genethics — The Ethics of Engineering Life,* Allen & Unwin, Sydney.
Thrupp, L (1998) *Cultivating Diversity: Agrobiodiversity and Food Security.* World Resources Institute, Washington DC.
Watson, JD (1970) *The Double Helix.* Penguin Books, London.
Weaver, WW (1997) *Heirloom Vegetable Gardening.* Henry Holt & Company, New York.
White, FWG (ed.) (1979) *Scientific Advances and Community Risk.* Australian Academy of Science, Canberra.

INDEX

Acacia 27
acid phosphatase 75
adenine 16, 18
Aequorea victoria 32
aflatoxin 134
Agent Orange 46
AgrEvo 61
agricultural practices
  sustainable 76, 120–25
  unsustainable 87, 92, 117, 120
Agrobacterium
  plasmids 37–38
  use in transformation 35, 37–38, 62–63, 66, 75, 81
ailanthone 53
alanine 50, 60
alleles
  dominant 24–25, 49, 79, 142
  recessive 24–25, 79, 111, 142, plates 1, 2
aluminium, toxicity 67
amaranth 123
amber, and DNA preservation 27–28
amino acids
  essential 50–51, 80, 133, 142
  polypeptide-forming 18, 142
ammonia 87, 120
ammonium ions 47, 120, 145
amylase inhibitor 62–63
Anabaena azollae 121
annual ryegrass 51
Anredera cordifolia 50
anther
  culture 39
  development 35

antisense sequences 34, 62, 64, 66
aphids 63, 119
apple 22, 59, 66
Arabidopsis thaliana 26, 66
Archer, Michael 28
Aristotle 23
armyworm 62–63
asparagine 50
Aspergillus niger 68, 94, 134
AstraZeneca 61, 70
Australia and New Zealand Food
  Authority 91, 95, 131
Australian Cultivar Registration
  Authority 103, 111
auxin 46
Aventis 61, 63, 90, 130
azafenidin 49, 53

Bacillus
  B. amyloliquefaciens 69
  B. thuringiensis 33, 63–64, 92 *see also* Bt-protein, Di Pel
Bailey, LH 102
banana 26, 68–69, 76, 81–82, 123
Banksia plate 11
Banks, Sir Joseph 22
barley 21, 47, 49, 51, 54
barnase 35–37, 69–70, 85, 92 *see also* Terminator
Bateson, William 23
bean
  broad 21, 49, 110, 121
  common 21, 51, 62, 121, 133, plate 19
biodiversity

advantages 122
  loss 87, 90, 117
  preservation 55
biolistics 35, 37
black gram 121
Borlaug, Norman 116
Bounty 107–109
Brassica 124
  B. campestris 21, 52
  B. napus 21–22 *see also* canola
  B. oleracea 21
Brazil 116
bread wheat *see Triticum aestivum*
Brown, Robert 14
Bt-proteins 33–34, 63–64, 85, 90–92, 120, 129, 133 *see also* INGARD
Burbank, Betty 101
Burbank, Luther 23, 59, 100–101

cadmium 50, 134
caffeine 81
calcium 67, 76
Callosobruchus maculatus 62
Cambodia 124
CaMV35S promoter 38, 63, 75, 93, 129
Canada 52, 86–87
canola 51, 66, 116, 130, plate 5
  composition 77, 91, 95 *see also* oils, canola
  gene transfer 45, 52, 94
  herbicide resistance 45, 52
capsicum 21, 134
  cultivation 134
  dried products 134
carnations 34
b-carotene 74, 76
carrot 21, 76, 123
cauliflower mosaic virus (CaMV) 93
cecropin 66
celery 110
cell division *see* meiosis, mitosis
cell wall 14, 19, 34, 66
Chakravarty, Anan Mohan 101
chickpea 21, 80, 116, 121, 133
chimeric gene constructs 32, 36, 38, 75
China 40, 115, 121, 124
chinese cabbage 134
chlorates 45–46
chlorophyll 53, 60, 70, 143
chloroplast 14–15
  genome 19–20, 35, 46
  origin 20
  proteins 20, 46
chlorsulfuron 47, 49, 50

chromosome
  number 21–22
  structure 21
Ciba Geigy 61
citrate, secretion by roots 67
citrus 53, 135
Clarke, Adrienne 86
clovers 55, 65, 67–68, 122
coat proteins, of viruses 65
codons 18
colchicine 39
competition, to exclude weeds 54–55
Consultative Group on International Agricultural Research Institutes 111, 116–17, 125
copy number, transferred genes 33, 35–37
corn earworm 62–63
Correa 103
cotton 66, 79, 81, 87, 117, 133, plate 7 *see also* INGARD
cotton bollworm 62
cowpea 21, 62
Crick, Francis 16
Crozier, Alan 81
CSIRO Horticulture 68
CSIRO Plant Industry 63, 65–66, 79, 81, 92, 96, 120, 131, 138, plates 6–8
CSIRO Tropical Agriculture 133
Cuba 122–24
Cucurbita 64
cultivation 54
cysteine 51, 80, 142
cytochromes 53, 143
cytokinin 46
cytoplasm 14–16, 143
cytoplasmic inheritance 16, 20
cytosine 16–17

2,4-D *see* 2,4-dinitrophenoxyacetic acid
daffodil 75
Danby, Michael 105
Dawson, Iain 103
DDT 49, 142
de Lamarck, Jean-Baptiste de Monet 14
diamondback moth 63
Diascia 104
2,4-dinitrophenoxyacetic acid 46–47, 49, 53
dioxins 48
Di Pel 92
dithiocarbamate 134
diuron 46

DNA
    complementary (cDNA) 143
    C-value 26, 143
    fragment patterns 26–28
    plastid 19–20
    preservation 27–28
    replication 17
    sequencing 26–28, 100
    structure 16–17
    synthesis 18
    transferred (T-DNA) 37–38
    viral 38, 147
Drosophila melanogaster 94
Du Pont 60–61
Dwyer, John 31

earthworms 46, 48
East Timor 124
Ecuador 124
eggplants 59, 125
Egypt 121
electroporation 34
elephant grass 121
embryo 20, 35
    rescue 40, 45
embryogenesis 20, 40, 143–44
    somatic 40–41, 45
endonucleases 18, 27, 38
endosperm 75, 77, 144
endosulfan 134
eosinophilia myalgia syndrome 32
erucic acid 77–78
Erwinia
    E. amylovora 66
    E. carotovora 66
    E. uredovora 75
Escherichia coli 32, 37
ethylene 34
eucalypts 53

Fanton, Jude 124–25
Fanton, Michel 124–25, plates 21, 27
Farrer, William 55
fatty acids, in plant oils 78
feeding trials, of transgenic plant products 63, 81, 88–91, 95, 131
fertilisers
    shortages 67
    wastage 87
fertilization 20, 22–25, 144
field trials
    new cultivars 103, 106
    transgenic plants 130
Fiji 124
fingerlimes 104–105
fireblight *see Erwinia amylovora*

firefly 32
fish antifreeze glycoprotein 60
Fisher, RA 24
Flinders, Matthew 14
fluorescence 32
food
    exports 50, 75, 91, 116, 134
    labelling problems 132
Food and Agriculture Organisation (FAO) 111, 125
Ford, Henry 100
frogs 48, 117
fungal disease resistance 20, 40, 66, 118 *see also* powdery mildew resistance

geminivirus 38
gene expression
    magnitude 33, 81–82
    tissue specificity 33–34, 38, 63, 67, 90, 93, 129
    see also microarray
gene probes 33
genetic diversity
    loss 87, 110, 117
    preservation 110, 119, 123–26
genetic engineering
    definition 38–40
    regulation 31, 39–40, 86, 92, 94–95, 130–32
    scope 40–41
genetic escape 44, 52, 87, 92, 94, 130
genetic uniformity 87, 118–19
gherkin 59
Gibbs, Adrian 93
Gluconacetobacter diazotrophicus 121
glucosinolates, in canola 77, 95
b-glucuronidase (GUS) 32–33
glufosinate 37, 47, 49, 51, 129
glycoproteins 60–61
glyphosate 46–50
    resistance 44, 52, 95
    seed content 50, 129
Goodrich, Chauncy 118
gossypol 81
grapes 67–68
grapevines 53, 66
Green, Dr Allan 79
Green Revolution 76, 116–18
Grevillea 103
Grew, Nehemiah 13
groundnut *see* peanut
guanine 16–18

Hartl, Dr Dan 94
Helicobacter pylori 93

Helminthosporium
 H. maydis 20, 118
 H. oryzae 117
Herbaspirillum 121
herbicides 44–55
 classification 47
 spray drift 45, 55, 87
 systemic 49–51
herbicide resistance
 in crop plants 44–45, 51–52 *see also* soybean, Roundup Ready
 in weeds 51
 and yield reduction 45
Heritage Seed Curators Australia 109, 111
Hippocrates 23–24
histone 16, 33
Hofmeister, Wilhelm 14, 23
Hooke, Robert 13
Hovey, Charles 23
Huebner, Leila 130
hybridisation 20–25, 35, 104, 111, 119
hypertension 81, 133

imidazolinone herbicides 47, 53
India 40, 75, 115, 121, 124–25, plates 25, 26
Indonesia 117
INGARD (cotton) 63, 92, 129, 134
intellectual property
 inherent contradictions 99–100
 detrimental impacts 109, 111–12
 see also plant breeders' rights, plant patents
International Union for the Protection of New Varieties of Plants 102, 110
iodosulfuron 53
irises 103
iron 76–77, 121
isoleucine 48, 50, 80, 142

Jurassic Park syndrome 27–28

Kaatz, Hans-Heinrich 94
kale 21
kanamycin 35
kangaroo paw 104, plate 10
Kenya 124
King, Philip Gidley 22
Knight, Thomas Andrew 22–23
Kuiper, Dr Harry 89

leafhoppers 38
lectins 61–63

from snowdrop 63, 88–90
legumes
 food value 62, 77, 80–81, 87, 116, 123
 nitrogen contribution 120–21
 phosphorus mobilisation 68, 120
 rotation with 54, 119–22, 135
lentil 133
lettuce 62
leucine 48, 50, 80, 142
ligase 18
locust 62
lucerne 20, 34, 40, 67
 crop loss 119
 weed contamination 54
lupin
 narrow-leafed 81, 121, plate 6
 white 67, 81
lysine 80, 142

magnesium 76, 143
maize 21, 26, 35, 40, 77, 82, 87, 90, 116, 129, plate 20 *see also* StarLink
malate, secretion by roots 67
Malawi 124
Malaysia 124, plates 18, 19
male sterility
 cytoplasmic 20
 nuclear 35–37
Malpighi, Marcello 13
mango 76, 134
marguerite daisy 104, plate 13
May, Sir Robert 89
Meek, Dr Sue 130
meiosis 22
Mendel, Gregor 23–25, 106
methionine 18, 51, 80–81, 142
methyl bromide 45
Mexico 116
microarray 95–96
microbodies 14–15, 20
milkweed 59, 90
millets 117, 121, 125, plate 26
mitochondrion 14–15, 19–20, 35
 genome 20, 143
 origin 20
mitosis 40, 62, 70
molecular taxonomy 26–28
monarch butterflies 90
Monsanto 33, 40–41, 44, 46, 61, 63, 70, 92, 116, 119, 125, 129–30
Morell, Dr Matthew 96
mung bean 121
mushrooms 133
mustard 124

mutagenesis 38–39
mutation 27, 39, 65, 93
mycorrhizae 48, 67

nectarine 111
nematodes 63
Nepal 124, plates 21–24
nicotinamide 35, 37
nitrate
  in food 120
  in soil 120, 145
nitrilase 66
nitrogen
  fertilisers 87, 120
  fixation 120–21
Novartis 61, 70

oats 46
Ochoa, Dr Carlos 117
oil seed rape 77 *see also* canola
oils
  canola 77–78
  cottonseed 79, 134
  linola 79
  macadamia 78
  maize 77
  olive 78, 133
  palm 79
  peanut 77–78
  sesame 77
  sunflower 79
Oken, Lorenz 14
okra 117
onion 21, 124
organic standards 135–36

papaya 76, 123
paper daisy 104, plates 16, 17
passionfruit 111
Pasteur, Louis 16
parthenium weed 54
Pauling, Linus 16
PBR *see* plant breeders' rights
pea
  field 62–63
  garden 21–26, 49–50, 102, 111,
    133, plates 1, 2 *see also* Bounty
peach 111
peanut 77–78, 121
pear 59
pea weevil 62
pesticides 50, 87, 93, 117–18, 133–35
petunia 104, plate 15
Phalaris paradoxa 55
Pharmacia & Upjohn 61
phenylalanine 46, 51, 80, 142

phloem 49–50, 145
phosphorus
  supply 67–68, 120
  unavailable 67–68, 120
  uptake 67
  see also phytate
photosynthesis 31, 120, 122
  inhibitors 46–47, 49
phyllode 27
phytase 68
phytate 68, 76–77
phyto-oestrogens 133
pigeon pea 121
pines 53
Pioneer Hi-Bred 45, 61
plant breeders' rights 100, 104–10,
  112
plant patents 75, 95–101
plasma membrane 14, 34
plasmids 37–38, 93, 101, 145
Plasmodiophora brassicae 66
plastics 82
pollen 14, 34
  germination 23
  production 35
  tube 23
pollination
  by humans 22, 25, 35
  by insects 52, 92, 130
  by wind 90
polyphenolics 68–69
polyphenoloxidase 68
polyploidy 21–22, 39
potassium 76–77
potato 27, 101, 133, plate 4
  breeding 63, 66, 88–90, 117–18
  late blight resistance 118
powdery mildew resistance 102, 107,
  111
prickly pear 45–46
promoters 32, 36–38, 40, 67, 69, 75,
  79, 92 *see also* CaMV35S promoter
protein
  bodies 76
  breakdown 70
  synthesis 18–20, 33, 69–70
proteinases 62
proteinase inhibitors 61–62
Protein Technologies International
  60–61
protoplasts
  fusion 40–41
  isolation 34
  plant regeneration 34, 41
  transformation 34, 38
Pseudomonas 68, 101

quarantine 66

radish 124
Ramprasad, Dr Vanaja 125
rats, in feeding trials 88–91, 95, 133
reporter genes 32–35, 37, 63, 94
Research Foundation for Science, Technology and Natural Resource Policy 125
Rhizobium leguminosarum 121
ribonuclease 18–19, 65, 79 *see also* barnase
ribosomes 14–15, 18, 20
ribozymes 65, 146
rice 21, 77
   breeding 35, 117, 124
   with b-carotene 74–76, 116
   cultivation 123, 135
   varieties 40, 124–25, 135, plates 25, 26
RNA
   messenger 18–19, 70
   ribosomal 18, 28, 70
   satellite 64–65
   transfer 18
   viral core 64–65
root systems 67–68, 120–21
roses 60, 103
Royal Horticultural Society 23, 103, 110
rye 21, 49, plate 9

Sabah 124
Salmonella typhimurium 46
Scaevola plate 14
Schleiden, Matthias 14, 23
Schmeiser, Percy 52
Schwann, Theodor 14
seed
   development 50, 69
   dispersal 51, 54, 130
   germination 38, 70, 92
   saving 102, 110, 118, 123–25, 135, plates 21–29
   storage 116, 125
seed proteins 33, 60, 69, 75, 79–81, 88 *see also* lectins, proteinase inhibitors
Seed Savers' Network 111, 124, 136, 140
Selden, George 100
serine 51
sesame 77
shasta daisy 104
Shiva, Dr Vandana 125
Showa-Denko 32, 86

silkworm 66
silverbeet 124
soil
   acidity 67
   bacteria 37, 53, 63, 68, 87, 117, 121–22
   erosion 87, 118
   fungi 48, 87
   weed seed bank 50, 54
solanine 89
Solomon Islands 124, plates 28, 29
somaclonal variation 40
somatic hybridisation 40
sorghum 21, 117
South Indian Seed Network 125
soybean 21, 49, 87–88
   breeding 119–20
   origin 119
   Roundup Ready 33, 44, 50, 116, 119, 129, 132–33
soy products, alternatives 133
spinach 124
Sri Lanka 122
Staphylococcus aureus 93
starch
   maize 82
   rice 77
   wheat 96
StarLink 63, 88, 90–91, 133
strawberry 22, 60, 134
stress 118–19, 122
Sturt's desert pea 104
sugar beet 49
sugar cane 40, 121
sugars 49, 63, 76, 120, 123, 133
sulfate 121
sulfonylurea herbicides 47–51, 53
Sumeghy, Joe 106
sunflower 79, 81
sweet potato 76, 123
Syngenta 61, 70, 94

2,4,5-T *see* 2,4,5-trinitrophenoxyacetic acid
Tasmania 107, 130
Tasmanian tiger 27–28
Terminator
   and male sterility 35–37
   and seeds 69–70, 118
tetracycline, as gene switch 69
tetraploids 21, 22, 39
Thailand 75
threonine 60, 80, 142
thymine 16–18
tissue culture 34, 38, 40–41, 45, 143–44

tobacco 62, 67
tobacco budworm 62–63, 81
tobacco hornworm 62–63
tomato 21, 26–27, 34, 40, 60, 65, 125
Tonga 124, plate 27
trade marks 110–11
transcription 18, 32, 147
transformation
  methodology 32–38
  rate 35, 37–38
transgene constructs 32, 36, 38, 75
translation 18, 147
tree of heaven 53
triazine herbicides 45–48
2,4,5-trinitrophenoxyacetic acid 46
Triticum 21–22
  T. aestivum 21–22, 35–38, 49, 66–67, 76–77, 87, 116, 135, plate 9
Trounce 107–109
tryptophan 18, 32, 51, 80, 142
turnip 21
tyrosine 46, 51, 80, 142

UNESCO 122
University of Sydney Plant Breeding Institute 96, 104, plates 9, 13–17
uracil 18

vacuole 14–15, 19
valine 48, 50, 80, 142
van Leewenhoek, Anton 13
Vibrio harveyi 32
Vietnam 76, 119, 122–23, plate 20
virus
  replication 64–65, 93
  resistance 40, 64–66
  transmission 64
vitamin A 74–76
vitamin E 77–78

waratah 104, 112, plate 12
Waterhouse, Dr Doug 104, 112
Watson, James 16–17
weed management 53–55
wheat *see Triticum*
whitefly 38
Whitmore, Bill 108–109
wild oats 54–55
Williams, Medwyn 110, 141

xylem 49, 66–67, 147

yams 123

Zimbabwe 124
zinc 76